Spectral Methods
in MATLAB

SOFTWARE • ENVIRONMENTS • TOOLS

The SIAM series on Software, Environments, and Tools focuses on the practical implementation of computational methods and the high performance aspects of scientific computation by emphasizing in-demand software, computing environments, and tools for computing. Software technology development issues such as current status, applications and algorithms, mathematical software, software tools, languages and compilers, computing environments, and visualization are presented.

Editor-in-Chief
Jack J. Dongarra
University of Tennessee and Oak Ridge National Laboratory

Editorial Board

James W. Demmel, University of California, Berkeley
Dennis Gannon, Indiana University
Eric Grosse, AT&T Bell Laboratories
Jorge J. Moré, Argonne National Laboratory

Software, Environments, and Tools

Jeremy Kepner, *Parallel MATLAB for Multicore and Multinode Computers*

Michael A. Heroux, Padma Raghavan, and Horst D. Simon, editors, *Parallel Processing for Scientific Computing*

Gérard Meurant, *The Lanczos and Conjugate Gradient Algorithms: From Theory to Finite Precision Computations*

Bo Einarsson, editor, *Accuracy and Reliability in Scientific Computing*

Michael W. Berry and Murray Browne, *Understanding Search Engines: Mathematical Modeling and Text Retrieval, Second Edition*

Craig C. Douglas, Gundolf Haase, and Ulrich Langer, *A Tutorial on Elliptic PDE Solvers and Their Parallelization*

Louis Komzsik, *The Lanczos Method: Evolution and Application*

Bard Ermentrout, *Simulating, Analyzing, and Animating Dynamical Systems: A Guide to XPPAUT for Researchers and Students*

V. A. Barker, L. S. Blackford, J. Dongarra, J. Du Croz, S. Hammarling, M. Marinova, J. Waśniewski, and P. Yalamov, *LAPACK95 Users' Guide*

Stefan Goedecker and Adolfy Hoisie, *Performance Optimization of Numerically Intensive Codes*

Zhaojun Bai, James Demmel, Jack Dongarra, Axel Ruhe, and Henk van der Vorst, *Templates for the Solution of Algebraic Eigenvalue Problems: A Practical Guide*

Lloyd N. Trefethen, *Spectral Methods in MATLAB*

E. Anderson, Z. Bai, C. Bischof, S. Blackford, J. Demmel, J. Dongarra, J. Du Croz, A. Greenbaum, S. Hammarling, A. McKenney, and D. Sorensen, *LAPACK Users' Guide, Third Edition*

Michael W. Berry and Murray Browne, *Understanding Search Engines: Mathematical Modeling and Text Retrieval*

Jack J. Dongarra, Iain S. Duff, Danny C. Sorensen, and Henk A. van der Vorst, *Numerical Linear Algebra for High-Performance Computers*

R. B. Lehoucq, D. C. Sorensen, and C. Yang, *ARPACK Users' Guide: Solution of Large-Scale Eigenvalue Problems with Implicitly Restarted Arnoldi Methods*

Randolph E. Bank, *PLTMG: A Software Package for Solving Elliptic Partial Differential Equations, Users' Guide 8.0*

L. S. Blackford, J. Choi, A. Cleary, E. D'Azevedo, J. Demmel, I. Dhillon, J. Dongarra, S. Hammarling, G. Henry, A. Petitet, K. Stanley, D. Walker, and R. C. Whaley, *ScaLAPACK Users' Guide*

Greg Astfalk, editor, *Applications on Advanced Architecture Computers*

Roger W. Hockney, *The Science of Computer Benchmarking*

Françoise Chaitin-Chatelin and Valérie Frayssé, *Lectures on Finite Precision Computations*

Spectral Methods in MATLAB

Lloyd N. Trefethen
Oxford University
Oxford, England

Society for Industrial and Applied Mathematics
Philadelphia

Copyright © 2000 by the Society for Industrial and Applied Mathematics.

10 9 8 7 6 5 4

All rights reserved. Printed in the United States of America. No part of this book may be reproduced, stored, or transmitted in any manner without the written permission of the publisher. For information, write to the Society for Industrial and Applied Mathematics, 3600 Market Street, 6th Floor, Philadelphia, PA 19104-2688 USA.

Trademarked names may be used in this book without the inclusion of a trademark symbol. These names are used in an editorial context only; no infringement of trademark is intended.

MATLAB is a registered trademark of The MathWorks, Inc. For MATLAB product information, please contact The MathWorks, Inc., 3 Apple Hill Drive, Natick, MA 01760-2098 USA, 508-647-7000, Fax: 508-647-7001, info@mathworks.com, www.mathworks.com.

Library of Congress Cataloging-in-Publication Data
Trefethen, Lloyd N. (Lloyd Nicholas)
    Spectral methods in MATLAB / Lloyd N. Trefethen.
       p. cm. – (Software, environments, tools)
    Includes bibliographical references and index.
    ISBN 978-0-898714-65-4 (pbk.)
       1. Differential equations, Partial—Numerical solutions—Data processing.
       2. Spectral theory (Mathematics) 3. MATLAB. I. Title. II. Series.

QA377.T65 2000
515'.7222–dc21

                                                    00-036559

 is a registered trademark.

To Anne

Contents

Preface	ix
Acknowledgments	xiii
A Note on the MATLAB Programs	xv
1 Differentiation Matrices	1
2 Unbounded Grids: The Semidiscrete Fourier Transform	9
3 Periodic Grids: The DFT and FFT	17
4 Smoothness and Spectral Accuracy	29
5 Polynomial Interpolation and Clustered Grids	41
6 Chebyshev Differentiation Matrices	51
7 Boundary Value Problems	61
8 Chebyshev Series and the FFT	75
9 Eigenvalues and Pseudospectra	87
10 Time-Stepping and Stability Regions	101
11 Polar Coordinates	115
12 Integrals and Quadrature Formulas	125
13 More about Boundary Conditions	135
14 Fourth-Order Problems	145
Afterword	153
Bibliography	155
Index	161

Preface

The aim of this book is to teach you the essentials of spectral collocation methods with the aid of 40 short MATLAB® programs, or "M-files."* The programs are available online at http://www.comlab.ox.ac.uk/oucl/work/nick.trefethen, and you will run them and modify them to solve all kinds of ordinary and partial differential equations (ODEs and PDEs) connected with problems in fluid mechanics, quantum mechanics, vibrations, linear and nonlinear waves, complex analysis, and other fields. Concerning prerequisites, it is assumed that the words just written have meaning for you, that you have some knowledge of numerical methods, and that you already know MATLAB.

If you like computing and numerical mathematics, you will enjoy working through this book, whether alone or in the classroom—and if you learn a few new tricks of MATLAB along the way, that's OK too!

Spectral methods are one of the "big three" technologies for the numerical solution of PDEs, which came into their own roughly in successive decades:

> 1950s: finite difference methods
> 1960s: finite element methods
> 1970s: spectral methods

Naturally, the origins of each technology can be traced further back. For spectral methods, some of the ideas are as old as interpolation and expan-

---

*MATLAB is a registered trademark of The MathWorks, Inc., 3 Apple Hill Drive, Natick, MA 01760-2098, USA, tel. 508-647-7000, fax 508-647-7001, info@mathworks.com, http://www.mathworks.com.

sion, and more specifically algorithmic developments arrived with Lanczos as early as 1938 [Lan38, Lan56] and with Clenshaw, Elliott, Fox, and others in the 1960s [FoPa68]. Then, in the 1970s, a transformation of the field was initiated by work by Orszag and others on problems in fluid dynamics and meteorology, and spectral methods became famous. Three landmarks of the early modern spectral methods literature were the short book by Gottlieb and Orszag [GoOr77], the survey by Gottlieb, Hussaini, and Orszag [GHO84], and the monograph by Canuto, Hussaini, Quarteroni, and Zang [CHQZ88]. Other books have been contributed since then by Mercier [Mer89], Boyd [Boy00] (first edition in 1989), Funaro [Fun92], Bernardi and Maday [BeMa92], Fornberg [For96], and Karniadakis and Sherwin [KaSh99].

If one wants to solve an ODE or PDE to high accuracy on a simple domain, and if the data defining the problem are smooth, then spectral methods are usually the best tool. They can often achieve ten digits of accuracy where a finite difference or finite element method would get two or three. At lower accuracies, they demand less computer memory than the alternatives.

This short textbook presents some of the fundamental ideas and techniques of spectral methods. It is aimed at anyone who has finished a numerical analysis course and is familiar with the basics of applied ODEs and PDEs. You will see that a remarkable range of problems can be solved to high precision by a few lines of MATLAB in a few seconds of computer time. Play with the programs; make them your own! The exercises at the end of each chapter will help get you started.

I would like to highlight three mathematical topics presented here that, while known to experts, are not usually found in textbooks. The first, in Chapter 4, is the connection between the smoothness of a function and the rate of decay of its Fourier transform, which determines the size of the aliasing errors introduced by discretization; these connections explain how the accuracy of spectral methods depends on the smoothness of the functions being approximated. The second, in Chapter 5, is the analogy between roots of polynomials and electric point charges in the plane, which leads to an explanation in terms of potential theory of why grids for nonperiodic spectral methods need to be clustered at boundaries. The third, in Chapter 8, is the three-way link between Chebyshev series on $[-1, 1]$, trigonometric series on $[-\pi, \pi]$, and Laurent series on the unit circle, which forms the basis of the technique of computing Chebyshev spectral derivatives via the fast Fourier transform. All three of these topics are beautiful mathematical subjects in their own right, well worth learning for any applied mathematician.

If you are determined to move immediately to applications without paying too much attention to the underlying mathematics, you may wish to turn directly to Chapter 6. Most of the applications appear in Chapters 7–14.

Inevitably, this book covers only a part of the subject of spectral methods. It emphasizes collocation ("pseudospectral") methods on periodic and on

Chebyshev grids, saying next to nothing about the equally important Galerkin methods and Legendre grids and polynomials. The theoretical analysis is very limited, and simple tools for simple geometries are emphasized rather than the "industrial strength" methods of spectral elements and hp finite elements. Some indications of omitted topics and other points of view are given in the Afterword.

A new era in scientific computing has been ushered in by the development of MATLAB. One can now present advanced numerical algorithms and solutions of nontrivial problems in complete detail with great brevity, covering more applied mathematics in a few pages than would have been imaginable a few years ago. By sacrificing sometimes (not always!) a certain factor in machine efficiency compared with lower level languages such as Fortran or C, one obtains with MATLAB a remarkable human efficiency—an ability to modify a program and try something new, then something new again, with unprecedented ease. This short book is offered as an encouragement to students, scientists, and engineers to become skilled at this new kind of computing.

Acknowledgments

I must begin by acknowledging two special colleagues who have taught me a great deal about spectral methods over the years. These are André Weideman, of the University of Stellenbosch, coauthor of the "MATLAB Differentiation Matrix Suite" [WeRe00], and Bengt Fornberg, of the University of Colorado, author of *A Practical Guide to Pseudospectral Methods* [For96]. These good friends share my enthusiasm for simplicity—and my enjoyment of the occasional detail that refuses to be simplified, no matter how hard you try. In this book, among many other contributions, Weideman significantly improved the crucial program `cheb`.

I must also thank Cleve Moler, the inventor of MATLAB, a friend and mentor since my graduate school days. Perhaps I had better admit here at the outset that there is a brass plaque on my office wall, given to me in 1998 by The MathWorks, Inc., which reads: *FIRST ORDER FOR MATLAB, February 7, 1985, Ordered by Professor Nick Trefethen, Massachusetts Institute of Technology.* I was there in the classroom at Stanford when Moler taught the numerical eigensystems course CS238b in the winter of 1979 based around this new experimental interface to EISPACK and LINPACK he had cooked up. I am a card-carrying MATLAB fan.

Toby Driscoll, author of the SC Toolbox for Schwarz–Christoffel conformal mapping in MATLAB [Dri96], has taught me many MATLAB tricks, and he helped to improve the codes in this book. He also provided key suggestions for the nonlinear waves calculations of Chapter 10. The other person whose suggestions improved the codes most significantly is Pascal Gahinet of The MathWorks, Inc., whose eye for MATLAB style is something special. David Carlisle of NAG, Ltd., one of the authors of LaTeX 2_ε, showed me how to make blank lines in MATLAB programs come out a little bit shortened when included as verbatim input, saving precious centimeters for display of figures. Walter Gautschi and Sotiris Notaris informed me about matters related to Clenshaw–Curtis quadrature, and Jean-Paul Berrut and Richard Baltensperger taught me about rounding errors in spectral differentiation matrices.

A number of other colleagues commented upon drafts of the book and improved it. I am especially grateful to John Boyd, Frédéric Dias, Des Higham, Nick Higham, Álvaro Meseguer, Paul Milewski, Damian Packer, and Satish Reddy.

In a category by himself goes Mark Embree, who has read this book more carefully than anyone else but me, by far. Embree suggested many improvements in the text, and beyond that, he worked many of the exercises, catching errors and contributing new exercises of his own. I am lucky to have found Embree at a stage of his career when he still has so much time to give to others.

The Numerical Analysis Group here at Oxford provides a stimulating environment to support a project like this. I want particularly to thank my three close colleagues Mike Giles, Endre Süli, and Andy Wathen, whose friendship has made me glad I came to Oxford; Shirley Dickson, who cheerfully made multiple copies of drafts of the text half a dozen times on short notice; and our outstanding group secretary and administrator, Shirley Day, who will forgive me, I hope, for all the mornings I spent working on this book when I should have been doing other things.

This book started out as a joint production with Andrew Spratley, a D. Phil. student, based on a masters-level course I taught in 1998 and 1999. I want to thank Spratley for writing the first draft of many of these pages and for major contributions to the book's layout and figures. Without his impetus, the book would not have been written.

Once we knew it would be written, there was no doubt who the publisher should be. It was a pleasure to publish my previous book [TrBa97] with SIAM, an organization that manages to combine the highest professional standards with a personal touch. And there was no doubt who the copy editor should be: again the remarkable Beth Gallagher, whose eagle eye and good sense have improved the presentation from beginning to end.

Finally, special thanks for their encouragement must go to my two favorite younger mathematicians, Emma (8) and Jacob (6) Trefethen, who know how I love differential equations, MATLAB, and writing. I'm the sort of writer who polishes successive drafts endlessly, and the children are used to seeing me sit down and cover a chapter with marks in red pen. Jacob likes to tease me and ask, "Did you find more mistakes in your book, Daddy?"

A Note on the MATLAB Programs

The MATLAB programs in this book are terse. I have tried to make each one compact enough to fit on a single page, and most often, on half a page. Of course, there is a message in this style, which is the message of this book: you can do an astonishing amount of serious computing in a few inches of computer code! And there is another message, too. The best discipline for making sure you understand something is to simplify it, simplify it relentlessly.

Without a doubt, readability is sometimes impaired by this obsession with compactness. For example, I have often combined two or three short MATLAB commands on a single program line. You may prefer a looser style, and that is fine. What's best for a printed book is not necessarily what's best for one's personal work.

Another idiosyncrasy of the programming style in this book is that the structure is flat: with the exception of the function `cheb`, defined in Chapter 6 and used repeatedly thereafter, I make almost no use of functions. (Three further functions, `chebfft`, `clencurt`, and `gauss`, are introduced in Chapters 8 and 12, but each is used just locally.) This style has the virtue of emphasizing how much can be achieved compactly, but as a general rule, MATLAB programmers should make regular use of functions.

Quite a bit might have been written to explain the details of each program, for there are tricks throughout this book that will be unfamiliar to some readers. To keep the discussion focused on spectral methods, I made a deliberate decision not to mention these MATLAB details except in a very few cases. This means that as you work with the book, you will have to study the programs, not just read them. What is this "`pol2cart`" command in Program 28

(p. 120)? What's going on with the index variable "b" in Program 36 (p. 142)? You will only understand the answers to questions like these after you have spent time with the codes and adapted them to solve your own problems. I think this is part of the fun of using this book, and I hope you agree.

The programs listed in these pages were included as M-files directly into the LaTeX source file, so all should run correctly as shown. The outputs displayed are exactly those produced by running the programs on my machine. There was a decision involved here. Did we really want to clutter the text with endless formatting and Handle Graphics commands such as `fontsize`, `markersize`, `subplot`, and `pbaspect`, which have nothing to do with the mathematics? In the end I decided that yes, we did. I want you to be able to download these programs and get beautiful results immediately. Equally important, experience has shown me that the formatting and graphics details of MATLAB are areas of this language where many users are particularly grateful for some help.

My personal MATLAB setup is nonstandard in one way: I have a file `startup.m` that contains the lines

```
set(0,'defaultaxesfontsize',12,'defaultaxeslinewidth',.7,...
    'defaultlinelinewidth',.8,'defaultpatchlinewidth',.7).
```

This makes text appear by default slightly larger than it otherwise would, and lines slightly thicker. The latter is important in preparing attractive output for a publisher's high-resolution printer.

The programs in this book were prepared using MATLAB versions 5.3 and 6.0. As later versions are released in upcoming years, unfortunately, it is possible that some difficulties with the programs will appear. Updated codes with appropriate modifications will be made available online as necessary.

To learn MATLAB from scratch, or for an outstanding reference, I recommend SIAM's new *MATLAB Guide*, by Higham and Higham [HiHi00].

Think globally. Act locally.

1. Differentiation Matrices

Our starting point is a basic question. Given a set of grid points $\{x_j\}$ and corresponding function values $\{u(x_j)\}$, how can we use this data to approximate the derivative of u? Probably the method that immediately springs to mind is some kind of finite difference formula. It is through finite differences that we shall motivate spectral methods.

To be specific, consider a uniform grid $\{x_1, \ldots, x_N\}$, with $x_{j+1} - x_j = h$ for each j, and a set of corresponding data values $\{u_1, \ldots, u_N\}$:

Let w_j denote the approximation to $u'(x_j)$, the derivative of u at x_j. The standard second-order finite difference approximation is

$$w_j = \frac{u_{j+1} - u_{j-1}}{2h}, \qquad (1.1)$$

which can be derived by considering the Taylor expansions of $u(x_{j+1})$ and $u(x_{j-1})$. For simplicity, let us assume that the problem is periodic and take $u_0 = u_N$ and $u_1 = u_{N+1}$. Then we can represent the discrete differentiation process as a matrix-vector multiplication,

$$\begin{pmatrix} w_1 \\ \vdots \\ w_N \end{pmatrix} = h^{-1} \begin{pmatrix} 0 & \frac{1}{2} & & & -\frac{1}{2} \\ -\frac{1}{2} & 0 & \ddots & & \\ & \ddots & \ddots & & \\ & & \ddots & 0 & \frac{1}{2} \\ \frac{1}{2} & & & -\frac{1}{2} & 0 \end{pmatrix} \begin{pmatrix} u_1 \\ \vdots \\ u_N \end{pmatrix}. \tag{1.2}$$

(Omitted entries here and in other sparse matrices in this book are zero.) Observe that this matrix is *Toeplitz*, having constant entries along diagonals; i.e., a_{ij} depends only on $i - j$. It is also *circulant*, meaning that a_{ij} depends only on $(i - j) \pmod{N}$. The diagonals "wrap around" the matrix.

An alternative way to derive (1.1) and (1.2) is by the following process of local interpolation and differentiation:

For $j = 1, 2, \ldots, N$:
- *Let p_j be the unique polynomial of degree ≤ 2 with $p_j(x_{j-1}) = u_{j-1}$, $p_j(x_j) = u_j$, and $p_j(x_{j+1}) = u_{j+1}$.*
- *Set $w_j = p'_j(x_j)$.*

It is easily seen that, for fixed j, the interpolant p_j is given by

$$p_j(x) = u_{j-1} a_{-1}(x) + u_j a_0(x) + u_{j+1} a_1(x),$$

where $a_{-1}(x) = (x - x_j)(x - x_{j+1})/2h^2$, $a_0(x) = -(x - x_{j-1})(x - x_{j+1})/h^2$, and $a_1(x) = (x - x_{j-1})(x - x_j)/2h^2$. Differentiating and evaluating at $x = x_j$ then gives (1.1).

This derivation by local interpolation makes it clear how we can generalize to higher orders. Here is the fourth-order analogue:

For $j = 1, 2, \ldots, N$:
- *Let p_j be the unique polynomial of degree ≤ 4 with $p_j(x_{j\pm 2}) = u_{j\pm 2}$, $p_j(x_{j\pm 1}) = u_{j\pm 1}$, and $p_j(x_j) = u_j$.*
- *Set $w_j = p'_j(x_j)$.*

Again assuming periodicity of the data, it can be shown that this prescription

1. Differentiation Matrices

amounts to the matrix-vector product

$$\begin{pmatrix} w_1 \\ \\ \vdots \\ \\ w_N \end{pmatrix} = h^{-1} \begin{pmatrix} \ddots & & & & \frac{1}{12} & -\frac{2}{3} \\ \ddots & -\frac{1}{12} & & & & \frac{1}{12} \\ \ddots & \frac{2}{3} & \ddots & & & \\ & \ddots & 0 & \ddots & & \\ & & \ddots & -\frac{2}{3} & \ddots & \\ & & & \frac{1}{12} & \ddots & \\ -\frac{1}{12} & & & \frac{2}{3} & -\frac{1}{12} & \ddots \end{pmatrix} \begin{pmatrix} u_1 \\ \\ \vdots \\ \\ u_N \end{pmatrix}. \quad (1.3)$$

This time we have a pentadiagonal instead of tridiagonal circulant matrix.

The matrices of (1.2) and (1.3) are examples of *differentiation matrices*. They have order of accuracy 2 and 4, respectively. That is, for data u_j obtained by sampling a sufficiently smooth function u, the corresponding discrete approximations to $u'(x_j)$ will converge at the rates $O(h^2)$ and $O(h^4)$ as $h \to 0$, respectively. One can verify this by considering Taylor series.

Our first MATLAB program, Program 1, illustrates the behavior of (1.3). We take $u(x) = e^{\sin(x)}$ to give periodic data on the domain $[-\pi, \pi]$:

$-\pi \quad x_1 \quad x_2 \hspace{8em} x_N = \pi$

The program compares the finite difference approximation w_j with the exact derivative, $e^{\sin(x_j)} \cos(x_j)$, for various values of N. Because it makes use of MATLAB sparse matrices, this code runs in a fraction of a second on a workstation, even though it manipulates matrices of dimensions as large as 4096 [GMS92]. The results are presented in Output 1, which plots the maximum error on the grid against N. The fourth-order accuracy is apparent. This is our first and last example that does not illustrate a spectral method!

We have looked at second- and fourth-order finite differences, and it is clear that consideration of sixth-, eighth-, and higher order schemes will lead to circulant matrices of increasing bandwidth. The idea behind spectral methods is to take this process to the limit, at least in principle, and work with a differentiation formula of infinite order and infinite bandwidth—i.e., a dense matrix [For75]. In the next chapter we shall show that in this limit, for an

Program 1

```
% p1.m - convergence of fourth-order finite differences
% For various N, set up grid in [-pi,pi] and function u(x):
  Nvec = 2.^(3:12);
  clf, subplot('position',[.1 .4 .8 .5])
  for N = Nvec
    h = 2*pi/N; x = -pi + (1:N)'*h;
    u = exp(sin(x)); uprime = cos(x).*u;

    % Construct sparse fourth-order differentiation matrix:
    e = ones(N,1);
    D =   sparse(1:N,[2:N 1],2*e/3,N,N)...
        - sparse(1:N,[3:N 1 2],e/12,N,N);
    D = (D-D')/h;

    % Plot max(abs(D*u-uprime)):
    error = norm(D*u-uprime,inf);
    loglog(N,error,'.','markersize',15), hold on
  end
  grid on, xlabel N, ylabel error
  title('Convergence of fourth-order finite differences')
  semilogy(Nvec,Nvec.^(-4),'--')
  text(105,5e-8,'N^{-4}','fontsize',18)
```

Output 1

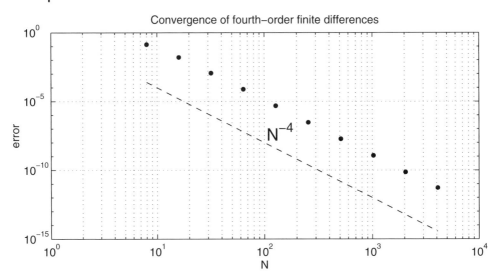

Output 1: *Fourth-order convergence of the finite difference differentiation process* (1.3). *The use of sparse matrices permits high values of N.*

1. Differentiation Matrices

infinite equispaced grid, one obtains the following infinite matrix:

$$D = h^{-1} \begin{pmatrix} & & & \vdots & & & \\ \ddots & & & \frac{1}{3} & & & \\ & \ddots & & -\frac{1}{2} & & & \\ & & \ddots & 1 & & & \\ & & & 0 & & & \\ & & & -1 & \ddots & & \\ & & & \frac{1}{2} & & \ddots & \\ & & & -\frac{1}{3} & & & \ddots \\ & & & \vdots & & & \end{pmatrix}. \qquad (1.4)$$

This is a skew-symmetric ($D^T = -D$) doubly infinite Toeplitz matrix, also known as a *Laurent operator* [Hal74, Wid65]. All its entries are nonzero except those on the main diagonal.

Of course, in practice one does not work with an infinite matrix. For a finite grid, here is the design principle for spectral collocation methods:

- Let p be a single function (independent of j) such that $p(x_j) = u_j$ for all j.
- Set $w_j = p'(x_j)$.

We are free to choose p to fit the problem at hand. For a periodic domain, the natural choice is a trigonometric polynomial on an equispaced grid, and the resulting "Fourier" methods will be our concern through Chapter 4 and intermittently in later chapters. For nonperiodic domains, algebraic polynomials on irregular grids are the right choice, and we will describe the "Chebyshev" methods of this type beginning in Chapters 5 and 6.

For finite N, taking N even for simplicity, here is the $N \times N$ dense matrix we will derive in Chapter 3 for a periodic, regular grid:

$$D_N = \begin{pmatrix} & & & \vdots & & & \\ \ddots & & & \frac{1}{2}\cot\frac{3h}{2} & & & \\ & \ddots & & -\frac{1}{2}\cot\frac{2h}{2} & & & \\ & & \ddots & \frac{1}{2}\cot\frac{1h}{2} & & & \\ & & & 0 & & & \\ & & & -\frac{1}{2}\cot\frac{1h}{2} & \ddots & & \\ & & & \frac{1}{2}\cot\frac{2h}{2} & & \ddots & \\ & & & -\frac{1}{2}\cot\frac{3h}{2} & & & \ddots \\ & & & \vdots & & & \end{pmatrix}. \qquad (1.5)$$

Program 2

```
% p2.m - convergence of periodic spectral method (compare p1.m)
% For various N (even), set up grid as before:
  clf, subplot('position',[.1 .4 .8 .5])
  for N = 2:2:100;
    h = 2*pi/N;
    x = -pi + (1:N)'*h;
    u = exp(sin(x)); uprime = cos(x).*u;

    % Construct spectral differentiation matrix:
    column = [0 .5*(-1).^(1:N-1).*cot((1:N-1)*h/2)];
    D = toeplitz(column,column([1 N:-1:2]));

    % Plot max(abs(D*u-uprime)):
    error = norm(D*u-uprime,inf);
    loglog(N,error,'.','markersize',15), hold on
  end
  grid on, xlabel N, ylabel error
  title('Convergence of spectral differentiation')
```

Output 2

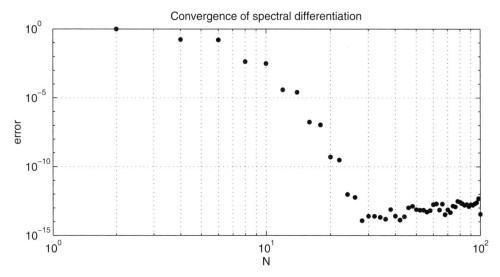

Output 2: *"Spectral accuracy" of the spectral method* (1.5), *until the rounding errors take over around* 10^{-14}. *Now the matrices are dense, but the values of N are much smaller than in Program 1.*

1. Differentiation Matrices

A little manipulation of the cotangent function reveals that this matrix is indeed circulant as well as Toeplitz (Exercise 1.2).

Program 2 is the same as Program 1 except with (1.3) replaced by (1.5). What a difference it makes in the results! The errors in Output 2 decrease very rapidly until such high precision is achieved that rounding errors on the computer prevent any further improvement.* This remarkable behavior is called *spectral accuracy*. We will give this phrase some precision in Chapter 4, but for the moment, the point to note is how different it is from convergence rates for finite difference and finite element methods. As N increases, the error in a finite difference or finite element scheme typically decreases like $O(N^{-m})$ for some constant m that depends on the order of approximation and the smoothness of the solution. For a spectral method, convergence at the rate $O(N^{-m})$ for *every* m is achieved, provided the solution is infinitely differentiable, and even faster convergence at a rate $O(c^N)$ ($0 < c < 1$) is achieved if the solution is suitably analytic.

The matrices we have described have been circulant. The action of a circulant matrix is a convolution, and as we shall see in Chapter 3, convolutions can be computed using a discrete Fourier transform (DFT). Historically, it was the discovery of the fast Fourier transform (FFT) for such problems in 1965 that led to the surge of interest in spectral methods in the 1970s. We shall see in Chapter 8 that the FFT is applicable not only to trigonometric polynomials on equispaced grids, but also to algebraic polynomials on Chebyshev grids. Yet spectral methods implemented without the FFT are powerful, too, and in many applications it is quite satisfactory to work with explicit matrices. Most problems in this book are solved via matrices.

Summary of This Chapter. The fundamental principle of spectral collocation methods is, given discrete data on a grid, to interpolate the data globally, then evaluate the derivative of the interpolant on the grid. For periodic problems, we normally use trigonometric interpolants in equispaced points, and for nonperiodic problems, we normally use polynomial interpolants in unevenly spaced points.

Exercises

1.1. We derived the entries of the tridiagonal circulant matrix (1.2) by local polynomial interpolation. Derive the entries of the pentadiagonal circulant matrix (1.3) in the same manner.

---

*All our calculations are done in standard IEEE double precision arithmetic with $\epsilon_{\text{machine}} = 2^{-53} \approx 1.11 \times 10^{-16}$. This means that each addition, multiplication, division, and subtraction produces the exactly correct result times some factor $1 + \delta$ with $|\delta| \leq \epsilon_{\text{machine}}$. See [Hig96] and [TrBa97].

1.2. Show that (1.5) is circulant.

1.3. The dots of Output 2 lie in pairs. Why? What property of $e^{\sin(x)}$ gives rise to this behavior?

1.4. Run Program 1 to $N = 2^{16}$ instead of 2^{12}. What happens to the plot of the error vs. N? Why? Use the MATLAB commands `tic` and `toc` to generate a plot of approximately how the computation time depends on N. Is the dependence linear, quadratic, or cubic?

1.5. Run Programs 1 and 2 with $e^{\sin(x)}$ replaced by (a) $e^{\sin^2(x)}$ and (b) $e^{\sin(x)|\sin(x)|}$ and with `uprime` adjusted appropriately. What rates of convergence do you observe? Comment.

1.6. By manipulating Taylor series, determine the constant C for an error expansion of (1.3) of the form $w_j - u'(x_j) \sim C h^4 u^{(5)}(x_j)$, where $u^{(5)}$ denotes the fifth derivative. Based on this value of C and on the formula for $u^{(5)}(x)$ with $u(x) = e^{\sin(x)}$, determine the leading term in the expansion for $w_j - u'(x_j)$ for $u(x) = e^{\sin(x)}$. (You will have to find $\max_{x \in [-\pi, \pi]} |u^{(5)}(x)|$ numerically.) Modify Program 1 so that it plots the dashed line corresponding to this leading term rather than just N^{-4}. This adjusted dashed line should fit the data almost perfectly. Plot the difference between the two on a log-log scale and verify that it shrinks at the rate $O(h^6)$.

2. Unbounded Grids: The Semidiscrete Fourier Transform

We now derive our first spectral method, as given by the doubly infinite matrix of (1.4). This scheme applies to a discrete, unbounded domain, so it is not a practical method. However, it does introduce the mathematical ideas needed for the derivation and analysis of the practical schemes we shall see later.

Our infinite grid is denoted by $h\mathbb{Z}$, with grid points $x_j = jh$ for $j \in \mathbb{Z}$, the set of all integers:

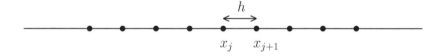

We shall derive (1.4) by various methods based on the key ideas of the semidiscrete Fourier transform and band-limited sinc function interpolation. Before discretizing, we review the continuous case [DyMc86, Kat76, Kör90]. The *Fourier transform* of a function $u(x)$, $x \in \mathbb{R}$, is the function $\hat{u}(k)$ defined by

$$\hat{u}(k) = \int_{-\infty}^{\infty} e^{-ikx} u(x)\, dx, \qquad k \in \mathbb{R. \tag{2.1}$$

The number $\hat{u}(k)$ can be interpreted as the amplitude density of u at wavenumber k, and this process of decomposing a function into its constituent waves is called *Fourier analysis*. Conversely, we can reconstruct u from $\hat{u}$ by the inverse Fourier transform:*

$$u(x) = \frac{1}{2\pi} \int_{-\infty}^{\infty} e^{ikx} \hat{u}(k)\, dk, \qquad x \in \mathbb{R}. \tag{2.2}$$

This is *Fourier synthesis*. The variable x is the *physical variable*, and k is the *Fourier variable* or *wavenumber*.

We want to consider x ranging over $h\mathbb{Z}$ rather than $\mathbb{R}$. Precise analogues of the Fourier transform and its inverse exist for this case. The crucial point is that because the spatial domain is discrete, the wavenumber k will no longer range over all of $\mathbb{R}$. Instead, the appropriate wavenumber domain is a bounded interval of length $2\pi/h$, and one suitable choice is $[-\pi/h, \pi/h]$. Remember, k is *bounded* because x is *discrete*:

$$\begin{array}{rcccl}
\text{Physical space} & : & \text{discrete, unbounded} & : & x \in h\mathbb{Z} \\
& & \updownarrow \qquad \updownarrow & & \\
\text{Fourier space} & : & \text{bounded, continuous} & : & k \in [-\pi/h, \pi/h]
\end{array}$$

The reason for these connections is the phenomenon known as *aliasing*. Two complex exponentials $f(x) = \exp(ik_1 x)$ and $g(x) = \exp(ik_2 x)$ are unequal over $\mathbb{R}$ if $k_1 \neq k_2$. If we restrict f and g to $h\mathbb{Z}$, however, they take values $f_j = \exp(ik_1 x_j)$ and $g_j = \exp(ik_2 x_j)$, and if $k_1 - k_2$ is an integer multiple of $2\pi/h$, then $f_j = g_j$ for each j. It follows that for any complex exponential $\exp(ikx)$, there are infinitely many other complex exponentials that match it on the grid $h\mathbb{Z}$ — "aliases" of k. Consequently it suffices to measure wavenumbers for the grid in an interval of length $2\pi/h$, and for reasons of symmetry, we choose the interval $[-\pi/h, \pi/h]$.

Figure 2.1 illustrates aliasing of the functions $\sin(\pi x)$ and $\sin(9\pi x)$. The dots represent restrictions to the grid $\frac{1}{4}\mathbb{Z}$, where these two functions are identical.

Aliasing occurs in nonmathematical life, too, for example in the "wagon wheel effect" in western movies. If, say, the shutter of a camera clicks 24 times a second and the spokes on a wagon wheel pass the vertical 20 times a second, then it looks as if the wheel is rotating at the rate of -4 spokes per second, i.e., backwards. Higher frequency analogues of the same phenomenon are the basis of the science of stroboscopy, and a spatial rather than temporal version of aliasing causes Moiré patterns.

For a function v defined on $h\mathbb{Z}$ with value v_j at x_j, the *semidiscrete Fourier transform* is defined by

$$\hat{v}(k) = h \sum_{j=-\infty}^{\infty} e^{-ikx_j} v_j, \qquad k \in [-\pi/h, \pi/h], \tag{2.3}$$

---

*Formulas (2.1) and (2.2) are valid, for example, for $u, \hat{u} \in L^2(\mathbb{R})$, the Hilbert space of complex square-integrable measurable functions on $\mathbb{R}$ [LiLo97]. However, this book will avoid most technicalities of measure theory and functional analysis.

2. Unbounded Grids: The Semidiscrete Fourier Transform

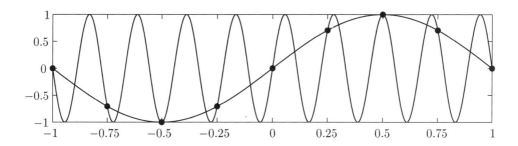

Fig. 2.1. *An example of aliasing. On the grid $\frac{1}{4}\mathbb{Z}$, the functions $\sin(\pi x)$ and $\sin(9\pi x)$ are identical.*

and the *inverse semidiscrete Fourier transform** is

$$v_j = \frac{1}{2\pi} \int_{-\pi/h}^{\pi/h} e^{ikx_j} \hat{v}(k)\, dk, \qquad j \in \mathbb{Z}. \tag{2.4}$$

Note that (2.3) approximates (2.1) by a trapezoid rule, and (2.4) approximates (2.2) by truncating $\mathbb{R}$ to $[-\pi/h, \pi/h]$. As $h \to 0$, the two pairs of formulas converge.

If the expression "semidiscrete Fourier transform" is unfamiliar, that may be because we have given a new name to an old concept. A *Fourier series* represents a function on a bounded interval as a sum of complex exponentials at discrete wavenumbers, as in (2.3). We have used the term semidiscrete Fourier transform to emphasize that our concern here is the inverse problem: it is the "space" variable that is discrete and the "Fourier" variable that is a bounded interval. Mathematically, there is no difference from the theory of Fourier series, which is presented in numerous books and is one of the most extensively worked branches of mathematics.

For spectral differentiation, we need an interpolant, and the inverse transform (2.4) will give us one. All we need to do is evaluate the same formula for $x \in \mathbb{R}$ rather than just $x_j \in h\mathbb{Z}$. That is, after determining $\hat{v}$, we define our interpolant p by

$$p(x) = \frac{1}{2\pi} \int_{-\pi/h}^{\pi/h} e^{ikx} \hat{v}(k)\, dk, \qquad x \in \mathbb{R}. \tag{2.5}$$

This is an analytic function of x,[†] with $p(x_j) = v_j$ for each j. Moreover, by

---

*These formulas hold for $v \in l^2(\mathbb{Z})$ (the set of square-summable grid functions) and $\hat{v} \in L^2[-\pi/h, \pi/h]$ (the set of square-integrable measurable functions on $[-\pi/h, \pi/h]$).

[†] A function f is *analytic* (or *holomorphic*) at a point $z \in \mathbb{C}$ if it is differentiable in the complex sense in a neighborhood of z, or equivalently, if its Taylor series converges to f

construction, the Fourier transform $\hat{p}$, defined by (2.1), is

$$\hat{p}(k) = \begin{cases} \hat{v}(k), & k \in [-\pi/h, \pi/h], \\ 0, & \text{otherwise.} \end{cases}$$

Thus $\hat{p}$ has compact support in $[-\pi/h, \pi/h]$. We say that p is the *band-limited interpolant* of v, by which we mean not just that $\hat{p}$ has compact support, but that this support is contained in the particular interval $[-\pi/h, \pi/h]$. Although there are an infinite number of possible interpolants for any grid function, there is only one band-limited interpolant defined in this sense. This result is known as the *sampling theorem* and is associated with the names of Whittaker, Shannon, and Nyquist [Hig85, OpSc89].

We are ready to give our first two descriptions of spectral differentiation of a function v defined on $h\mathbb{Z}$. Here is one:

- *Given v, determine its band-limited interpolant p by (2.5).*
- *Set $w_j = p'(x_j)$.*

Another is obtained by saying the same thing in Fourier space. If u is a differentiable function with Fourier transform $\hat{u}$, then the Fourier transform of u' is $ik\hat{u}(k)$:

$$\widehat{u'}(k) = ik\hat{u}(k). \tag{2.6}$$

This result can be obtained by differentiating (2.2) or (2.5) with respect to x. And thus we have an equivalent procedure for spectral differentiation:

- *Given v, compute its semidiscrete Fourier transform $\hat{v}$ by (2.3).*
- *Define $\hat{w}(k) = ik\hat{v}(k)$.*
- *Compute w from $\hat{w}$ by (2.4).*

Both of these descriptions of spectral differentiation are mathematically complete, but we have not yet derived the coefficients of the matrix (1.4). To do this, we can use the Fourier transform to go back and get a fuller understanding of the band-limited interpolant $p(x)$. Let δ be the *Kronecker delta function*,

$$\delta_j = \begin{cases} 1, & j = 0, \\ 0, & j \neq 0. \end{cases} \tag{2.7}$$

---

in a neighborhood of z [Ahl79, Hen74, Hil62]. In (2.5), $p(x)$ is analytic, for example, for $\hat{v} \in L^1[-\pi/h, \pi/h]$, hence in particular if $\hat{v}$ is in the smaller class $L^2[-\pi/h, \pi/h]$. This is equivalent to the condition $v \in \ell^2(\mathbb{Z})$.

2. Unbounded Grids: The Semidiscrete Fourier Transform

By (2.3), the semidiscrete Fourier transform of δ is a constant: $\hat{\delta}(k) = h$ for all $k \in [-\pi/h, \pi/h]$. By (2.5), the band-limited interpolant of δ is accordingly

$$p(x) = \frac{h}{2\pi} \int_{-\pi/h}^{\pi/h} e^{ikx} \, dk = \frac{\sin(\pi x/h)}{\pi x/h}$$

(with the value 1 at $x = 0$). This famous and beautiful function is called the *sinc function*,

$$S_h(x) = \frac{\sin(\pi x/h)}{\pi x/h}. \tag{2.8}$$

Sir Edmund Whittaker called S_1 "a function of royal blood in the family of entire functions, whose distinguished properties separate it from its bourgeois brethren" [Whi15]. For more about sinc functions and associated numerical methods, see [Ste93].

Now that we know how to interpolate the delta function, we can interpolate anything. Band-limited interpolation is a translation-invariant process in the sense that for any m, the band-limited interpolant of δ_{j-m} is $S_h(x - x_m)$. A general grid function v can be written

$$v_j = \sum_{m=-\infty}^{\infty} v_m \delta_{j-m}, \tag{2.9}$$

so it follows by the linearity of the semidiscrete Fourier transform that the band-limited interpolant of v is a linear combination of translated sinc functions:

$$p(x) = \sum_{m=-\infty}^{\infty} v_m S_h(x - x_m). \tag{2.10}$$

The derivative is accordingly

$$w_j = p'(x_j) = \sum_{m=-\infty}^{\infty} v_m S_h'(x_j - x_m). \tag{2.11}$$

And now let us derive the entries of the doubly infinite Toeplitz matrix D of (1.4). If we interpret (2.11) as a matrix equation as in (1.5), we see that the vector $S_h'(x_j)$ is the column $m = 0$ of D, with the other columns obtained by shifting this column up or down appropriately. The entries of (1.4) are determined by the calculus exercise of differentiating (2.8) to get

$$S_h'(x_j) = \begin{cases} 0, & j = 0, \\ \dfrac{(-1)^j}{jh}, & j \neq 0. \end{cases} \tag{2.12}$$

Program 3

```
% p3.m - band-limited interpolation
  h = 1; xmax = 10; clf
  x = -xmax:h:xmax;                    % computational grid
  xx = -xmax-h/20:h/10:xmax+h/20;      % plotting grid
  for plt = 1:3
    subplot(4,1,plt)
    switch plt
      case 1, v = (x==0);              % delta function
      case 2, v = (abs(x)<=3);         % square wave
      case 3, v = max(0,1-abs(x)/3);   % hat function
    end
    plot(x,v,'.','markersize',14), grid on
    p = zeros(size(xx));
    for i = 1:length(x),
      p = p + v(i)*sin(pi*(xx-x(i))/h)./(pi*(xx-x(i))/h);
    end
    line(xx,p), axis([-xmax xmax -.5 1.5])
    set(gca,'xtick',[]), set(gca,'ytick',[0 1])
  end
```

Output 3

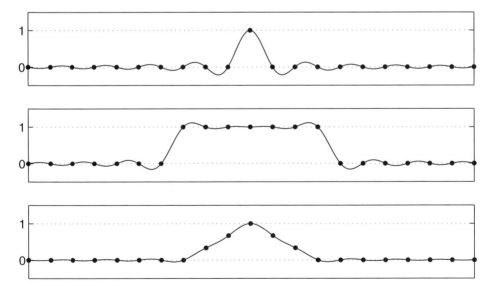

Output 3: *Band-limited interpolation of three grid functions; the first interpolant is the sinc function $S_h(x)$. Such interpolants are the basis of spectral methods, but these examples are not smooth enough for high accuracy.*

2. Unbounded Grids: The Semidiscrete Fourier Transform

A similar pattern applies to all spectral collocation methods. The mth column of a spectral differentiation matrix contains the values $p'(x_j)$, where $p(x)$ is the global interpolant through the discrete delta function supported at x_m.

Program 3 illustrates band-limited interpolation by plotting three discrete functions defined on $\mathbb{Z}$ (i.e., $h = 1$) together with their band-limited interpolants. It is clear from the first of the three plots in Output 3 that the sinc function is smooth. The second plot, depicting the interpolant of a discrete square wave, shows that sinc interpolants are not particularly good for approximating nonsmooth functions. The oscillations near the discontinuity will not diminish in amplitude as $h \to 0$, and they are not even well localized in space. This generation of oscillations near discontinuities is called the *Gibbs phenomenon*. The third plot shows a discrete triangular wave or "hat function" and its interpolant. Here the interpolation is somewhat better, but it is still not impressive. In fact, as we will explain in detail in Chapter 4, the accuracy of the interpolation depends upon the smoothness of u, and these examples are not very smooth. Each extra derivative possessed by u improves the order of accuracy by 1.

To find higher order spectral derivatives, we can differentiate $p(x)$ several times. For example, the result

$$S_h''(x_j) = \begin{cases} -\dfrac{\pi^2}{3h^2}, & j = 0, \\ 2\dfrac{(-1)^{j+1}}{j^2 h^2}, & j \neq 0, \end{cases} \tag{2.13}$$

tells us the entries of each column of the symmetric doubly infinite Toeplitz matrix D^2 corresponding to the second derivative:

$$D^2 v = \frac{2}{h^2} \begin{pmatrix} \ddots & & & \vdots & & & \\ & \ddots & & -\frac{1}{4} & & & \\ & & \ddots & 1 & & & \\ & & & -\frac{\pi^2}{6} & & & \\ & & & 1 & \ddots & & \\ & & & -\frac{1}{4} & & \ddots & \\ & & & \vdots & & & \ddots \end{pmatrix} v. \tag{2.14}$$

Summary of This Chapter. A function v on the grid $h\mathbb{Z}$ has a unique interpolant p that is band-limited to wavenumbers in the interval $[-\pi/h, \pi/h]$. We can compute p' on the grid by evaluating the inverse semidiscrete Fourier transform of $ik\hat{v}$, or alternatively, as a linear combination of derivatives of translates of sinc functions.

Exercises

2.1. Let $\mathcal{F}$ denote the Fourier transform operator defined by (2.1), so that $u, v \in L^2(\mathbb{R})$ have Fourier transforms $\hat{u} = \mathcal{F}\{u\}$, $\hat{v} = \mathcal{F}\{v\}$. Verify the following properties. Do not worry about rigorously justifying operations on integrals.

(a) *Linearity.* $\mathcal{F}\{u+v\}(k) = \hat{u}(k) + \hat{v}(k)$; $\mathcal{F}\{cu\}(k) = c\hat{u}(k)$.
(b) *Translation.* If $x_0 \in \mathbb{R}$, then $\mathcal{F}\{u(x+x_0)\}(k) = e^{ikx_0}\hat{u}(k)$.
(c) *Modulation.* If $k_0 \in \mathbb{R}$, then $\mathcal{F}\{e^{ik_0 x}u(x)\}(k) = \hat{u}(k-k_0)$.
(d) *Dilation.* If $c \in \mathbb{R}$ with $c \neq 0$, then $\mathcal{F}\{u(cx)\}(k) = \hat{u}(k/c)/|c|$.
(e) *Conjugation.* $\mathcal{F}\{\overline{u}\}(k) = \overline{\hat{u}(-k)}$.
(f) *Differentiation.* If $u_x \in L^2(\mathbb{R})$, then $\mathcal{F}\{u_x\}(k) = ik\hat{u}(k)$.
(g) *Inversion.* $\mathcal{F}^{-1}\{u\}(k) = (2\pi)^{-1}\hat{u}(-k)$.

2.2. Let $u \in L^2(\mathbb{R})$ have Fourier transform $\hat{u}$. Verify the following identities. (A function $f(x)$ is *hermitian* (*skew-hermitian*) if $f(-x) = \overline{f(x)}$ ($f(-x) = -\overline{f(x)}$).)

(a) $u(x)$ is even (odd) $\iff \hat{u}(k)$ is even (odd);
(b) $u(x)$ is real (imaginary) $\iff \hat{u}(k)$ is hermitian (skew-hermitian);
(c) $u(x)$ is real and even $\iff \hat{u}(k)$ is real and even;
(d) $u(x)$ is real and odd $\iff \hat{u}(k)$ is imaginary and odd;
(e) $u(x)$ is imaginary and even $\iff \hat{u}(k)$ is imaginary and even;
(f) $u(x)$ is imaginary and odd $\iff \hat{u}(k)$ is real and odd.

2.3. Execute the command `plot(sin(1:3000),'.')` in MATLAB. What do you see? What does this have to do with aliasing? Give a quantitative answer, explaining exactly what frequency is being aliased by your eye and brain to what other frequency. Then, for fun, replace `3000` by `1000` to get a figure somewhat harder to explain. (This problem comes from [Str91].)

2.4. Derive (2.12) and (2.13).

2.5. The text states that the matrix of (2.14) is the square of that of (1.4). Since these are infinite matrices, this amounts to an infinite set of assertions that certain series sum to certain values. What series and what values, exactly? In particular, what series sums to $\pi^2/6$? Sketch the argument by which, in the text and Exercise 2.4, we have implicitly calculated the sum of this series, which is famous as the value $\zeta(2)$ of the Riemann zeta function.

2.6. We obtained the entries of (1.4) by differentiating the sinc function. Derive the same result by calculating the inverse semidiscrete Fourier transform of $ik\hat{\delta}(k)$.

2.7. Modify Program 3 to determine the maximum error over $\mathbb{R}$ in the sinc function interpolants of the square wave and the hat function, and to produce a log-log plot of these two error maxima as functions of h. (Good choices for h are $2^{-3}, 2^{-4}, 2^{-5}, 2^{-6}$.) What convergence rates do you observe as $h \to 0$?

2.8. Differentiation of $u(x) = e^{ikx}$ multiplies it by $g_\infty(k) = ik$. Determine the analogous functions $g_2(k)$ and $g_4(k)$ corresponding to the finite difference processes (1.2) and (1.3), and draw a plot of $g_2(k)$, $g_4(k)$, and $g_\infty(k)$ vs. k. Where in the plot do we see the order of accuracy of a finite difference formula? (See [For75, Tre82].)

3. Periodic Grids: The DFT and FFT

We now turn to spectral differentiation on a bounded, periodic grid. This process was stated in the form of an $N \times N$ matrix operation in equation (1.5). Whereas the last chapter dealt with the infinite matrix (1.4) corresponding to the unbounded grid $h\mathbb{Z}$, this chapter develops a practical scheme for computation. Mathematically, there are close connections between the two schemes, and the derivation of our spectral method will follow the same pattern as before. The difference is that the semidiscrete Fourier transform is replaced by the DFT (discrete Fourier transform), which can be computed by the FFT (Fast Fourier Transform) [BrHe95].

At first sight, the requirement of periodicity may suggest that this method has limited relevance for practical problems. Yet periodic grids are surprisingly useful in practice. Often in scientific computing a phenomenon is of interest that is unrelated to boundaries, such as the interaction of solitons in the Korteweg–de Vries (KdV) equation (see p. 112) or the behavior of homogeneous turbulence (a classic reference is [OrPa72]). For such problems, periodic boundary conditions often prove the best choice for computation. In addition, some geometries are physically periodic, such as crystal lattices or rows of turbine blades. Finally, even if the physics is not periodic, the coordinate space may be, as is the case for a θ or ϕ variable in a computation involving polar or spherical coordinates (see Chapter 11). Indeed, among the most common of all spectral methods are methods that mix periodic grids in one or two angular directions with nonperiodic grids in one or two radial or longitudinal directions.

Our basic periodic grid will be a subset of the interval $[0, 2\pi]$:

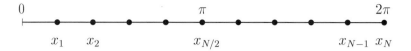

(Intervals of lengths other than 2π are easily handled by scale factors, and translations to other intervals such as $[-\pi, \pi]$ make no difference at all.) When we talk about a periodic grid, we mean that any data values on the grid come from evaluating a periodic function. Equivalently, we may regard the periodic grid as one cycle of length N extracted from an infinite grid with data satisfying $v_{j+mN} = v_j$ for all $j, m \in \mathbb{Z}$.

Throughout the book, the number of grid points on a periodic grid will always be even:

$$N \text{ is even.}$$

All our results have analogues for odd N, but the formulas are different, and little would be gained by writing everything twice. The spacing of the grid points is $h = 2\pi/N$, which gives us

$$\frac{\pi}{h} = \frac{N}{2}. \tag{3.1}$$

We recommend that the reader memorize this equation. The quotient π/h will appear over and over, since $[-\pi/h, \pi/h]$ is the range of wavenumbers distinguishable on the grid. We will make constant use of (3.1).

Let us consider the Fourier transform on the N-point grid. As in the last chapter, the mesh spacing h implies that wavenumbers differing by an integer multiple of $2\pi/h$ are indistinguishable on the grid, and thus it will be enough to confine our attention to $k \in [-\pi/h, \pi/h]$. Now, however, the Fourier domain is discrete as well as bounded. This is because waves in physical space must be periodic over the interval $[0, 2\pi]$, and only waves e^{ikx} with integer wavenumbers have the required period 2π. We find:

Physical space : discrete, bounded : $x \in \{h, 2h, \ldots, 2\pi - h, 2\pi\}$
$\updownarrow \quad\quad\quad \updownarrow$
Fourier space : bounded, discrete : $k \in \{-\frac{N}{2}+1, -\frac{N}{2}+2, \ldots, \frac{N}{2}\}$

The formula for the DFT is

$$\hat{v}_k = h \sum_{j=1}^{N} e^{-ikx_j} v_j, \qquad k = -\frac{N}{2}+1, \ldots, \frac{N}{2}, \tag{3.2}$$

3. Periodic Grids: The DFT and FFT

and the *inverse DFT* is given by

$$v_j = \frac{1}{2\pi} \sum_{k=-N/2+1}^{N/2} e^{ikx_j} \hat{v}_k, \qquad j = 1, \ldots, N. \tag{3.3}$$

These formulas may be compared with the formulas (2.1)–(2.2) for the Fourier transform and its inverse and (2.3)–(2.4) for the semidiscrete Fourier transform and its inverse.

In (3.2) and (3.3), the wavenumber k, like the spatial index j, takes only integer values. Nothing continuous or infinite is left in the problem, and (3.2) and (3.3) are inverses of one another for arbitrary vectors $(v_1, \ldots, v_N)^T \in \mathbf{C}^N$, without any technical restrictions.

For spectral differentiation of a grid function v, we follow the model of the last chapter exactly. First we need a band-limited interpolant of v, which we obtain by evaluating the inverse DFT (3.3) for all x rather than just x on the grid. But before we differentiate our interpolant, there is a complication to address, suggested in Figure 3.1. Evaluating the inverse transform (3.3) as it stands would give a term $e^{iNx/2}$ with derivative $(iN/2)e^{iNx/2}$. Since $e^{iNx/2}$ represents a real, sawtooth wave on the grid, its derivative should be zero at the grid points, not a complex exponential! The problem is that (3.3) treats the highest wavenumber asymmetrically. We can fix this by defining $\hat{v}_{-N/2} = \hat{v}_{N/2}$ and replacing (3.3) by

$$v_j = \frac{1}{2\pi} \sideset{}{'}\sum_{k=-N/2}^{N/2} e^{ikx_j} \hat{v}_k, \qquad j = 1, \ldots, N, \tag{3.4}$$

where the prime indicates that the terms $k = \pm N/2$ are multiplied by $\frac{1}{2}$. Note that the DFT and its inverse can still be defined by (3.2) and (3.3). Equation (3.4) is needed just for the purpose of deriving a band-limited interpolant, which takes the form

$$p(x) = \frac{1}{2\pi} \sideset{}{'}\sum_{k=-N/2}^{N/2} e^{ikx} \hat{v}_k, \qquad x \in [0, 2\pi], \tag{3.5}$$

and the sawtooth now has derivative zero, as it should. Note that $p(x)$ is a *trigonometric polynomial* of degree at most $N/2$. This means that it can be written as a linear combination of the functions $1, \sin(x), \cos(x), \sin(2x), \ldots, \sin(Nx/2), \cos(Nx/2)$. (Actually, the $\sin(Nx/2)$ term is not needed.)

To interpolate a grid function v, as in the previous chapter, we can use this general formula, or we can compute the band-limited interpolant of the delta function and expand v as a linear combination of translated delta functions. The delta function is now periodic (compare (2.7)):

$$\delta_j = \begin{cases} 1, & j \equiv 0 \pmod{N}, \\ 0, & j \not\equiv 0 \pmod{N}. \end{cases} \tag{3.6}$$

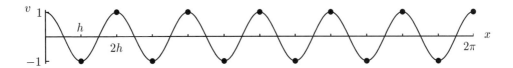

Fig. 3.1. *The sawtooth grid function and its interpolant* $\cos(\pi x/h) = \frac{1}{2}(e^{i\pi x/h} + e^{-i\pi x/h})$ *of* (3.5). *The appropriate spectral derivative is zero at every grid point, which is not the result we would get if we used the interpolant* $e^{i\pi x/h}$.

From (3.2) we have $\hat{\delta}_k = h$ for each k, and so, from (3.5), we get

$$p(x) = \frac{h}{2\pi} \sum_{k=-N/2}^{N/2}{}' e^{ikx} = \frac{h}{2\pi} \left(\frac{1}{2} \sum_{k=-N/2}^{N/2-1} e^{ikx} + \frac{1}{2} \sum_{k=-N/2+1}^{N/2} e^{ikx} \right)$$

$$= \frac{h}{2\pi} \cos(x/2) \sum_{k=-N/2+1/2}^{N/2-1/2} e^{ikx}$$

$$= \frac{h}{2\pi} \cos(x/2) \frac{e^{i(-N/2+1/2)x} - e^{i(N/2+1/2)x}}{1 - e^{ix}}$$

$$= \frac{h}{2\pi} \cos(x/2) \frac{e^{-i(N/2)x} - e^{i(N/2)x}}{e^{-ix/2} - e^{ix/2}}$$

$$= \frac{h}{2\pi} \cos(x/2) \frac{\sin(Nx/2)}{\sin(x/2)}.$$

Using the identity (3.1), we conclude that the band-limited interpolant to δ is the *periodic sinc function* S_N (Figure 3.2):

$$S_N(x) = \frac{\sin(\pi x/h)}{(2\pi/h)\tan(x/2)}. \tag{3.7}$$

Note that since $\tan x/2 \sim x/2$ as $x \to 0$, $S_N(x)$ behaves like the nonperiodic sinc function $S_h(x)$ of (2.8) in the limit $x \to 0$—independently of h.

An expansion of a periodic grid function v in the basis of shifted periodic delta functions takes the form $v_j = \sum_{m=1}^{N} v_m \delta_{j-m}$, in analogy to (2.9). Thus the band-limited interpolant of (3.5) can be written in analogy to (2.10) as

$$p(x) = \sum_{m=1}^{N} v_m S_N(x - x_m). \tag{3.8}$$

3. Periodic Grids: The DFT and FFT

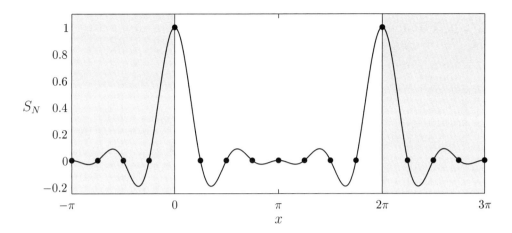

Fig. 3.2. *The periodic sinc function S_N, the band-limited interpolant of the periodic delta function δ, plotted for $N = 8$.*

After a small calculation analogous to the derivation of (2.12), we find that the derivative of the periodic sinc function at a grid point is

$$S'_N(x_j) = \begin{cases} 0, & j \equiv 0 \pmod{N}, \\ \frac{1}{2}(-1)^j \cot(jh/2), & j \not\equiv 0 \pmod{N}. \end{cases} \quad (3.9)$$

As discussed in the last chapter, these numbers are the entries of one column of the spectral differentiation matrix—the Nth column of the $N \times N$ matrix, since $N \equiv 0 \pmod{N}$. This was the matrix presented already in (1.5):

$$D_N = \begin{pmatrix} 0 & & & & -\frac{1}{2}\cot\frac{1h}{2} \\ -\frac{1}{2}\cot\frac{1h}{2} & \ddots & & \ddots & \frac{1}{2}\cot\frac{2h}{2} \\ \frac{1}{2}\cot\frac{2h}{2} & & \ddots & & -\frac{1}{2}\cot\frac{3h}{2} \\ -\frac{1}{2}\cot\frac{3h}{2} & & & \ddots & \vdots \\ \vdots & & \ddots & \ddots & \frac{1}{2}\cot\frac{1h}{2} \\ \frac{1}{2}\cot\frac{1h}{2} & & & & 0 \end{pmatrix}. \quad (3.10)$$

Program 4 illustrates the use of D_N for spectral differentiation of a hat function and the smooth function $e^{\sin(x)}$ of Programs 1 and 2. The accuracy for the hat function is poor, because the function is not smooth, but the accuracy for $e^{\sin(x)}$ is outstanding.

To calculate higher spectral derivatives, we can differentiate p, the inter-

Program 4

```
% p4.m - periodic spectral differentiation

% Set up grid and differentiation matrix:
  N = 24; h = 2*pi/N; x = h*(1:N)';
  column = [0 .5*(-1).^(1:N-1).*cot((1:N-1)*h/2)]';
  D = toeplitz(column,column([1 N:-1:2]));

% Differentiation of a hat function:
  v = max(0,1-abs(x-pi)/2); clf
  subplot(3,2,1), plot(x,v,'.-','markersize',13)
  axis([0 2*pi -.5 1.5]), grid on, title('function')
  subplot(3,2,2), plot(x,D*v,'.-','markersize',13)
  axis([0 2*pi -1 1]), grid on, title('spectral derivative')

% Differentiation of exp(sin(x)):
  v = exp(sin(x)); vprime = cos(x).*v;
  subplot(3,2,3), plot(x,v,'.-','markersize',13)
  axis([0 2*pi 0 3]), grid on
  subplot(3,2,4), plot(x,D*v,'.-','markersize',13)
  axis([0 2*pi -2 2]), grid on
  error = norm(D*v-vprime,inf);
  text(2.2,1.4,['max error = ' num2str(error)])
```

Output 4

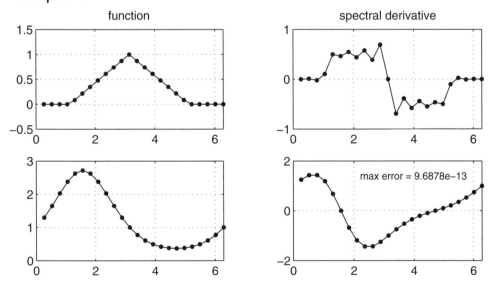

Output 4: *Spectral differentiation of a rough function and a smooth one. The smooth function gives* 12-*digit accuracy.*

3. Periodic Grids: The DFT and FFT

polant of (3.8), several times. Some calculus gives

$$S_N''(x_j) = \begin{cases} -\dfrac{\pi^2}{3h^2} - \dfrac{1}{6}, & j \equiv 0 \pmod{N}, \\ -\dfrac{(-1)^j}{2\sin^2(jh/2)}, & j \not\equiv 0 \pmod{N}. \end{cases} \quad (3.11)$$

Thus second-order spectral differentiation can be written in the matrix form

$$D_N^{(2)} v = \begin{pmatrix} \ddots & & \vdots & & \\ & \ddots & -\frac{1}{2}\csc^2(\frac{2h}{2}) & & \\ & & \frac{1}{2}\csc^2(\frac{1h}{2}) & & \\ & & -\frac{\pi^2}{3h^2} - \frac{1}{6} & & \\ & & \frac{1}{2}\csc^2(\frac{1h}{2}) & \ddots & \\ & & -\frac{1}{2}\csc^2(\frac{2h}{2}) & & \ddots \\ & & \vdots & & \ddots \end{pmatrix} v. \quad (3.12)$$

We can construct the matrix for differentiation of any order by further differentiation of S_N.

As ever with spectral methods, alternative methods of implementation can be found. In particular, the DFT can be used:

- *Given v, compute $\hat{v}$.*
- *Define $\hat{w}_k = ik\hat{v}_k$, except $\hat{w}_{N/2} = 0$.*
- *Compute w from $\hat{w}$.*

For higher derivatives we multiply by the appropriate power of ik, taking special care of the $\hat{w}_{N/2}$ term. For odd derivatives there is a loss of symmetry and we have to set $\hat{w}_{N/2} = 0$. Otherwise $\hat{w}_{N/2}$ is given by the same formula as the other $\hat{w}_k$. In summary, to approximate the νth derivative,

- *Given v, compute $\hat{v}$.*
- *Define $\hat{w}_k = (ik)^\nu \hat{v}_k$, but $\hat{w}_{N/2} = 0$ if ν is odd.*
- *Compute w from $\hat{w}$.*

The computation of the DFT can be accomplished by the FFT, discovered in 1965 by Cooley and Tukey. (Actually, the FFT was discovered by Gauss at the age of 28 in 1805—two years before Fourier completed his first big article!—but although Gauss wrote a paper on the subject, he did not publish it, and it lay unnoticed [HJB85].) If N is highly composite, that is, a product of small prime factors, then the FFT enables us to compute the DFT, and hence spectral derivatives, in $O(N \log N)$ floating point operations.

Program 5 is a repetition of Program 4 based on the FFT instead of matrices. The transforms are calculated by MATLAB's programs `fft` and `ifft`. A little care has to be exercised in using these programs since MATLAB assumes a different ordering of the wavenumbers from ours. In our notation, MATLAB stores wavenumbers in the order $0, 1, \ldots, \frac{N}{2}, -\frac{N}{2}+1, -\frac{N}{2}+2, \ldots, -1$.

There is an inefficiency in the use of the FFT as in Program 5 that must be mentioned. In most applications, the data v to be differentiated will be real, and yet, the use of the FFT makes use of a complex transform. Mathematically, the transform satisfies the symmetry property $\hat{v}(-k) = \overline{\hat{v}(k)}$ (Exercise 2.2), but MATLAB's standard FFT routines are not equipped to take advantage of this property. In an implementation in Fortran or C, one would expect to gain at least a factor of 2 by doing so. A different trick to recover that factor of 2 is considered in Exercise 3.6.

We have now developed spectral tools that can be applied in practice. To close the chapter, we will use the FFT method to solve a partial differential equation (PDE). Consider the variable coefficient wave equation

$$u_t + c(x)u_x = 0, \qquad c(x) = \tfrac{1}{5} + \sin^2(x-1) \tag{3.13}$$

for $x \in [0, 2\pi]$, $t > 0$, with periodic boundary conditions. As an initial condition we take $u(x,0) = \exp(-100(x-1)^2)$. This function is not mathematically periodic, but it is so close to zero at the ends of the interval that it can be regarded as periodic in practice.

To construct our numerical scheme, we proceed just as we might with a finite difference approximation of a PDE. For the time derivative we use a leap frog formula [Ise96], and we approximate the spatial derivative spectrally. Let $v^{(n)}$ be the vector at time step n that approximates $u(x_j, n\Delta t)$. At grid point x_j, our spectral derivative is $(Dv^{(n)})_j$, where $D = D_N$, the spectral differentiation matrix (1.5) for the N-point equispaced grid, which we implement by the FFT. The approximation becomes

$$\frac{v_j^{(n+1)} - v_j^{(n-1)}}{2\Delta t} = -c(x_j)(Dv^{(n)})_j, \qquad j = 1, \ldots, N.$$

This formula is all there is to Program 6, except for the complication that the leap frog scheme requires two initial conditions to start, whereas the PDE provides only one. To obtain a starting value $v^{(-1)}$, this particular program extrapolates backwards with the assumption of a constant wave speed of $\tfrac{1}{5}$. This approximation introduces a small error. For more serious work one could use one or more steps of a one-step ordinary differential equation (ODE) formula, such as a Runge–Kutta formula, to generate the necessary second set of initial data at $t = -\Delta t$ or $t = \Delta t$.

3. Periodic Grids: The DFT and FFT

Program 5

```
% p5.m - repetition of p4.m via FFT
%        For complex v, delete "real" commands.

% Differentiation of a hat function:
  N = 24; h = 2*pi/N; x = h*(1:N)';
  v = max(0,1-abs(x-pi)/2); v_hat = fft(v);
  w_hat = 1i*[0:N/2-1 0 -N/2+1:-1]' .* v_hat;
  w = real(ifft(w_hat)); clf
  subplot(3,2,1), plot(x,v,'.-','markersize',13)
  axis([0 2*pi -.5 1.5]), grid on, title('function')
  subplot(3,2,2), plot(x,w,'.-','markersize',13)
  axis([0 2*pi -1 1]), grid on, title('spectral derivative')

% Differentiation of exp(sin(x)):
  v = exp(sin(x)); vprime = cos(x).*v;
  v_hat = fft(v);
  w_hat = 1i*[0:N/2-1 0 -N/2+1:-1]' .* v_hat;
  w = real(ifft(w_hat));
  subplot(3,2,3), plot(x,v,'.-','markersize',13)
  axis([0 2*pi 0 3]), grid on
  subplot(3,2,4), plot(x,w,'.-','markersize',13)
  axis([0 2*pi -2 2]), grid on
  error = norm(w-vprime,inf);
  text(2.2,1.4,['max error = ' num2str(error)])
```

Output 5

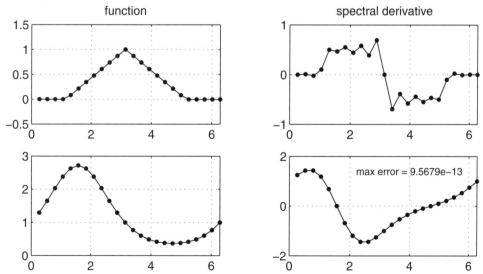

Output 5: *Repetition of Output 4 based on FFT instead of matrices.*

Program 6

```
% p6.m - variable coefficient wave equation

% Grid, variable coefficient, and initial data:
  N = 128; h = 2*pi/N; x = h*(1:N); t = 0; dt = h/4;
  c = .2 + sin(x-1).^2;
  v = exp(-100*(x-1).^2); vold = exp(-100*(x-.2*dt-1).^2);

% Time-stepping by leap frog formula:
  tmax = 8; tplot = .15; clf, drawnow
  plotgap = round(tplot/dt); dt = tplot/plotgap;
  nplots = round(tmax/tplot);
  data = [v; zeros(nplots,N)]; tdata = t;
  for i = 1:nplots
    for n = 1:plotgap
      t = t+dt;
      v_hat = fft(v);
      w_hat = 1i*[0:N/2-1 0 -N/2+1:-1] .* v_hat;
      w = real(ifft(w_hat));
      vnew = vold - 2*dt*c.*w; vold = v; v = vnew;
    end
    data(i+1,:) = v; tdata = [tdata; t];
  end
  waterfall(x,tdata,data), view(10,70), colormap([0 0 0])
  axis([0 2*pi 0 tmax 0 5]), ylabel t, zlabel u, grid off
```

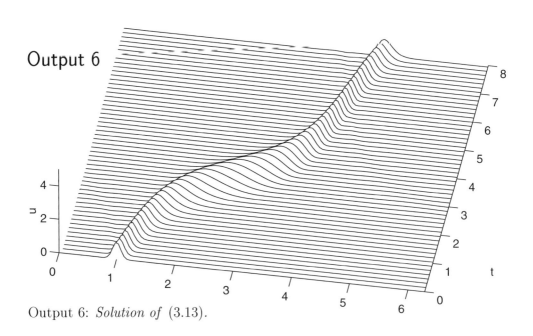

Output 6: *Solution of* (3.13).

3. Periodic Grids: The DFT and FFT

Output 6 shows a beautiful wave propagating at variable speed. There are no ripples emanating from the wave, which remains coherent and clean. Solutions from finite difference discretizations rarely look so nice (Exercise 3.9). The absence of spurious dispersion is one of the conspicuous advantages of spectral methods.

Summary of This Chapter. Mathematically, a periodic grid is much like an infinite grid, with the semidiscrete Fourier transform replaced by the DFT and the sinc function S_h replaced by the periodic sinc function S_N. The band-limited interpolant of a grid function is a trigonometric polynomial. Spectral derivatives can be calculated by a differentiation matrix in $O(N^2)$ or by the FFT in $O(N \log N)$ floating point operations.

Exercises

3.1. Determine D_N, D_N^2, and $D_N^{(2)}$ for $N = 2$ and 4, and confirm that in both cases, $D_N^2 \neq D_N^{(2)}$.

3.2. Derive (3.9) and (3.11).

3.3. Derive formulas for the entries of the third-order periodic Fourier spectral differentiation matrix $D_N^{(3)}$.

3.4. The errors printed in the bottom-right figures of Outputs 4 and 5 differ by about 1.5%. In the light of Output 2, explain this number. Why do they not agree exactly? Given that they disagree, why is it only in the second decimal place? How would your answers change if we took $N = 20$? $N = 30$?

3.5. Using the commands `tic` and `toc`, study the time taken by MATLAB (which is far from optimal in this regard) to compute an N-point FFT, as a function of N. If $N = 2^k$ for $k = 0, 1, \ldots, 15$, for example, what is the dependence of the time on N? What if $N = 500, 501, \ldots, 519, 520$? From a plot of the latter results, can you spot the prime numbers in this range? (*Hints.* The commands `isprime` and `factor` may be useful. To get good `tic`/`toc` data it may help to compute each FFT 10 or 100 times in a loop.)

3.6. We have seen that a discrete function v can be spectrally differentiated by means of two complex FFTs (one forward, one inverse). Explain how two distinct discrete functions v and w can be spectrally differentiated at once by the same two complex FFTs, provided that v and w are real.

3.7. Recompute Output 6 by a modified program based on matrices rather than the FFT. How much slower or faster is it? (Measure just the computation time, not the time for the `waterfall` plot.) How does the answer change if N is increased from 128 to 256?

3.8. In continuation of Exercise 3.7, modify your matrix program further so that instead of a leap frog discretization, it uses an explicit matrix exponential (MATLAB's `expm` command) to march directly from one plotting time step to the next.

How does this affect the computation time? How does it affect the accuracy?

3.9. Recompute Output 6 by a modified program based on the finite difference leap frog formula

$$\frac{v_j^{(n+1)} - v_j^{(n-1)}}{2\Delta t} = -c(x_j)\frac{v_{j+1}^{(n)} - v_{j-1}^{(n)}}{2\Delta x}, \qquad j = 1, \ldots, N,$$

rather than a spectral method. Produce plots for $N = 128$ and 256. Comment.

3.10. The solution of (3.13) is periodic in time: for a certain $T \approx 13$, $u(x,T) = u(x,0)$ for all x. Determine T analytically by evaluating an appropriate integral. Then modify Program 6 to compute $u(x,T)$ instead of $u(x,8)$. (Make sure t stops exactly at T, not at some nearby number determined by **round**.) For $N = 32, 64, \ldots, 512$, determine $\max_j |u(x_j,T) - u(x_j,0)|$, and plot this error on a log-log scale as a function of N. What is the rate of convergence? How could it be improved?

4. Smoothness and Spectral Accuracy

We are ready to discuss the accuracy of spectral methods. As stated in Chapter 1, the typical convergence rate is $O(N^{-m})$ for every m for functions that are smooth (fast!) and $O(c^N)$ ($0 < c < 1$) for functions that are analytic (faster!). Such behavior is known as *spectral accuracy*.

To derive these relationships we shall make use of the Fourier transform in an argument consisting of two steps. First, a smooth function has a rapidly decaying transform. The reason is that a smooth function changes slowly, and since high wavenumbers correspond to rapidly oscillating waves, such a function contains little energy at high wavenumbers. Second, if the Fourier transform of a function decays rapidly, then the errors introduced by discretization are small. The reason is that these errors are caused by aliasing of high wavenumbers to low wavenumbers.

We carry out the argument for the real line $\mathbb{R}$; similar reasoning applies in the periodic case. The following theorem collects four statements relating smoothness of u and decay of $\hat{u}$. Each condition on the smoothness of u is stronger than the last and implies a correspondingly faster decay rate for $\hat{u}$. This theorem makes use of standard mathematical ideas (differentiability, analyticity, complex plane) that some readers may be less familiar with than they would like. Rest assured at least that the book does not get more technical than this! The "big O" symbol is used with its usual precise meaning: $f(k) = O(g(k))$ as $k \to \infty$ if there exists a constant C such that for all sufficiently large k, $|f(k)| < C|g(k)|$. Similarly, the "little O" symbol is defined in the standard way: $f(k) = o(g(k))$ as $k \to \infty$ if $\lim_{k \to \infty} |f(k)|/|g(k)| = 0$.

We shall not prove Theorem 1, but refer the reader to the literature. Part

(a) can be handled by standard methods of analysis, and part (b) is a corollary. Parts (c) and (d) are known as *Paley–Wiener theorems*; see [Kat76] and [PaWe34].

Theorem 1 Smoothness of a function and decay of its Fourier transform.

Let $u \in L^2(\mathbb{R})$ have Fourier transform $\hat{u}$.

(a) *If u has $p-1$ continuous derivatives in $L^2(\mathbb{R})$ for some $p \geq 0$ and a pth derivative of bounded variation,* then*

$$\hat{u}(k) = O(|k|^{-p-1}) \qquad \text{as } |k| \to \infty.$$

(b) *If u has infinitely many continuous derivatives in $L^2(\mathbb{R})$, then*

$$\hat{u}(k) = O(|k|^{-m}) \qquad \text{as } |k| \to \infty$$

for every $m \geq 0$. The converse also holds.

(c) *If there exist $a, c > 0$ such that u can be extended to an analytic function in the complex strip $|\operatorname{Im} z| < a$ with $\|u(\cdot + iy)\| \leq c$ uniformly for all $y \in (-a, a)$, where $\|u(\cdot + iy)\|$ is the L^2 norm along the horizontal line $\operatorname{Im} z = y$, then $u_a \in L^2(\mathbb{R})$, where $u_a(k) = e^{a|k|}\hat{u}(k)$. The converse also holds.*

(d) *If u can be extended to an entire function (i.e., analytic throughout the complex plane) and there exists $a > 0$ such that $|u(z)| = o(e^{a|z|})$ as $|z| \to \infty$ for all complex values $z \in \mathbb{C}$, then $\hat{u}$ has compact support contained in $[-a, a]$; that is,*

$$\hat{u}(k) = 0 \qquad \text{for all } |k| > a.$$

The converse also holds.

This theorem, although technical, can be illustrated by examples of functions that satisfy the various smoothness conditions.

Illustration of Theorem 1(a). Consider the step function $s(x)$ defined by

$$s(x) = \begin{cases} \frac{1}{2}, & |x| \leq 1, \\ 0, & |x| > 1. \end{cases}$$

---

*A function f has *bounded variation* if it belongs to $L^1(\mathbb{R})$ and the supremum of $\int fg'$ over all $g \in C^1(\mathbb{R})$ with $|g(x)| \leq 1$ is finite [Zie89]. If f is continuous, this coincides with the condition that the supremum of $\sum_{j=1}^{N} |f(x_j) - f(x_{j-1})|$ over all finite samples $x_0 < x_1 < \cdots < x_N$ is bounded.

4. Smoothness and Spectral Accuracy

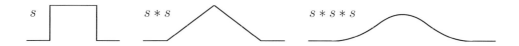

Fig. 4.1. *A step function s is used to generate piecewise polynomials—B-splines—by convolution.*

This function is not differentiable, but it has bounded variation. To generate functions with finite numbers of continuous derivatives, we can take *convolutions* of s with itself. The convolution of two functions u and v is defined by

$$(u * v)(x) = \int_{-\infty}^{\infty} u(y) v(x-y) \, dy = \int_{-\infty}^{\infty} v(y) u(x-y) \, dy, \qquad (4.1)$$

and the functions s, $s*s$, and $s*s*s$ are sketched in Figure 4.1. These functions, known as *B-splines*, are piecewise polynomials that vanish outside the intervals $[-1, 1]$, $[-2, 2]$, and $[-3, 3]$, respectively [Boo78].

The function s is piecewise constant and satisfies the condition of Theorem 1(a) with $p = 0$. Similarly, $s*s$ is piecewise linear, satisfying the condition with $p = 1$, and $s*s*s$ is piecewise quadratic, satisfying it with $p = 2$. Now the Fourier transforms of these functions are

$$\hat{s}(k) = \frac{\sin k}{k}, \quad \widehat{s*s}(k) = \left(\frac{\sin k}{k}\right)^2, \quad \widehat{s*s*s}(k) = \left(\frac{\sin k}{k}\right)^3.$$

These results follow from the general formula for the Fourier transform of a convolution,

$$\widehat{u * v}(k) = \hat{u}(k) \hat{v}(k), \qquad k \in \mathbb{R}. \qquad (4.2)$$

Just as predicted by Theorem 1(a), these three Fourier transforms decay at the rates $O(|k|^{-1})$, $O(|k|^{-2})$, and $O(|k|^{-3})$.

Illustration of Theorem 1(d). By reversing the roles of $\hat{s}$, $\widehat{s*s}$, $\widehat{s*s*s}$ and s, $s*s$, $s*s*s$ in the above example, that is, by regarding the former as the functions and the latter as the transforms (apart from some unimportant constant factors), we obtain illustrations of Theorem 1(d). The function $\hat{s}$, for example, satisfies $\hat{s}(k) = o(e^{|k|})$ as $|k| \to \infty$, and its Fourier transform, $2\pi s(x)$, has compact support $[-1, 1]$.

Illustration of Theorem 1(c). Consider the pair

$$u(x) = \frac{\sigma}{x^2 + \sigma^2}, \qquad \hat{u}(k) = \pi e^{-\sigma |k|} \qquad (4.3)$$

for any constant $\sigma > 0$. This function $u(x)$ is analytic throughout the complex plane except for poles at $\pm i\sigma$. Thus Theorem 1(c) applies, and we may take any $a < \sigma$. (The condition $\|u(\cdot + iy)\| \le c$ will fail if we take $a = \sigma$.) As predicted, the Fourier transform decays exponentially at the corresponding rate. By reversing the roles of u and $\hat{u}$ in this example, we obtain a function satisfying condition (a) of the theorem with $p = 1$ whose transform decays at the predicted rate $O(|k|^{-2})$.

Another illustration. For a final illustration let us consider that most familiar of all Fourier transform pairs,

$$u(x) = e^{-x^2/2\sigma^2}, \qquad \hat{u}(k) = \sigma\sqrt{\frac{\pi}{2}}\, e^{-\sigma^2 k^2/2}.$$

These functions fit between parts (c) and (d) of Theorem 1. A Gaussian pulse is "smoother" than analytic in a strip, but "less smooth" than entire with a growth condition at ∞; it is entire, but fails the growth condition since the exponential growth is quadratic rather than linear along rays other than the real axis. Correspondingly, its Fourier transform, another Gaussian pulse, decays faster than exponentially, since the exponent is squared, but does not have compact support.

Theorem 1 quantifies our first argument mentioned at the beginning of the chapter: a smooth function has a rapidly decaying Fourier transform. We now turn to the second argument, the connection between the decay of the Fourier transform and accuracy of the band-limited interpolant. For simplicity, we deal with the case of interpolation on the infinite grid $h\mathbb{Z}$. The essential idea for periodic grids is the same.

Let $u \in L^2(\mathbb{R})$ have a first derivative of bounded variation. This is (more than) enough to imply, by Theorem 1(a), that the series we are about to consider converge. Let v be the grid function obtained by sampling u on the grid $h\mathbb{Z}$. We ask, what is the relationship between $\hat{u}$, the Fourier transform of u, and $\hat{v}$, the semidiscrete Fourier transform of v? From our discussion of aliasing, it should be clear that for $k \in [-\pi/h, \pi/h]$, $\hat{v}(k)$ will consist of $\hat{u}(k)$ plus the sum of all additional components of wavenumbers that are indistinguishable from k on the grid. The following theorem to this effect is known as the *aliasing formula* or the *Poisson summation formula*.

Theorem 2 Aliasing formula.

Let $u \in L^2(\mathbb{R})$ have a first derivative of bounded variation, and let v be the grid function on $h\mathbb{Z}$ defined by $v_j = u(x_j)$. Then for all $k \in [-\pi/h, \pi/h]$,

$$\hat{v}(k) = \sum_{j=-\infty}^{\infty} \hat{u}(k + 2\pi j/h). \qquad (4.4)$$

4. Smoothness and Spectral Accuracy

From Theorem 2 we deduce that

$$\hat{v}(k) - \hat{u}(k) = \sum_{\substack{j=-\infty \\ j\neq 0}}^{\infty} \hat{u}(k + 2\pi j/h),$$

and so, combining this result with Theorem 1, we may relate the smoothness of u to the error $\hat{v}(k) - \hat{u}(k)$. If u is smooth, then $\hat{u}$ and $\hat{v}$ are close. The following theorem is readily derived from Theorems 1 and 2 (Exercise 4.1). For more extensive treatments along these lines, see [Hen86] and [Tad86].

Theorem 3 *Effect of discretization on the Fourier transform.*

Let $u \in L^2(\mathbb{R})$ have a first derivative of bounded variation, and let v be the grid function on $h\mathbb{Z}$ defined by $v_j = u(x_j)$. The following estimates hold uniformly for all $k \in [-\pi/h, \pi/h]$, or a fortiori, for $k \in [-A, A]$ for any constant A.

(a) *If u has $p - 1$ continuous derivatives in $L^2(\mathbb{R})$ for some $p \geq 1$ and a pth derivative of bounded variation, then*

$$|\hat{v}(k) - \hat{u}(k)| = O(h^{p+1}) \quad \text{as } h \to 0.$$

(b) *If u has infinitely many continuous derivatives in $L^2(\mathbb{R})$, then*

$$|\hat{v}(k) - \hat{u}(k)| = O(h^m) \quad \text{as } h \to 0$$

for every $m \geq 0$.

(c) *If there exist $a, c > 0$ such that u can be extended to an analytic function in the complex strip $|\text{Im } z| < a$ with $\|u(\cdot + iy)\| \leq c$ uniformly for all $y \in (-a, a)$, then*

$$|\hat{v}(k) - \hat{u}(k)| = O(e^{-\pi(a-\epsilon)/h}) \quad \text{as } h \to 0$$

for every $\epsilon > 0$.

(d) *If u can be extended to an entire function and there exists $a > 0$ such that for $z \in \mathbb{C}$, $|u(z)| = o(e^{a|z|})$ as $|z| \to \infty$, then, provided $h \leq \pi/a$,*

$$\hat{v}(k) = \hat{u}(k).$$

In various senses, Theorem 3 makes precise the notion of spectral accuracy of $\hat{v}$ as an approximation to $\hat{u}$. By *Parseval's identity* we have, in the appropriately defined 2-norms,

$$\|\hat{u}\| = \sqrt{2\pi}\|u\|, \quad \|\hat{v}\| = \sqrt{2\pi}\|v\|. \tag{4.5}$$

It follows that the functions defined by the inverse Fourier transform over $[-\pi/h, \pi/h]$ of $\hat{u}$ and $\hat{v}$ are also spectrally close in agreement. The latter function is nothing else than p, the band-limited interpolant of v. Thus from Theorem 3 we can readily derive bounds on $\|u(x) - p(x)\|_2$, and with a little more work, on pointwise errors $|u(x) - p(x)|$ in function values or $|u^{(\nu)}(x) - p^{(\nu)}(x)|$ in the νth derivative (Exercise 4.2).

Theorem 4 *Accuracy of Fourier spectral differentiation.*

Let $u \in L^2(\mathbb{R})$ have a νth derivative ($\nu \geq 1$) of bounded variation, and let w be the νth spectral derivative of u on the grid $h\mathbb{Z}$. The following estimates hold uniformly for all $x \in h\mathbb{Z}$.

(a) *If u has $p-1$ continuous derivatives in $L^2(\mathbb{R})$ for some $p \geq \nu + 1$ and a pth derivative of bounded variation, then*

$$|w_j - u^{(\nu)}(x_j)| = O(h^{p-\nu}) \quad \text{as } h \to 0.$$

(b) *If u has infinitely many continuous derivatives in $L^2(\mathbb{R})$, then*

$$|w_j - u^{(\nu)}(x_j)| = O(h^m) \quad \text{as } h \to 0$$

for every $m \geq 0$.

(c) *If there exist $a, c > 0$ such that u can be extended to an analytic function in the complex strip $|\operatorname{Im} z| < a$ with $\|u(\cdot + iy)\| \leq c$ uniformly for all $y \in (-a, a)$, then*

$$|w_j - u^{(\nu)}(x_j)| = O(e^{-\pi(a-\epsilon)/h}) \quad \text{as } h \to 0$$

for every $\epsilon > 0$.

(d) *If u can be extended to an entire function and there exists $a > 0$ such that for $z \in \mathbb{C}$, $|u(z)| = o(e^{a|z|})$ as $|z| \to \infty$, then, provided $h \leq \pi/a$,*

$$w_j = u^{(\nu)}(x_j).$$

Program 7 illustrates the various convergence rates we have discussed. The program computes the spectral derivatives of four periodic functions, $|\sin x|^3$, $\exp(-\sin^{-2}(x/2))$, $1/(1 + \sin^2(x/2))$, and $\sin(10x)$. The first has a third derivative of bounded variation, the second is smooth but not analytic, the third is analytic in a strip in the complex plane, and the fourth is band-limited. The ∞-norm of the error in the spectral derivative is calculated for various step sizes, and in Output 7 we see varying convergence rates as predicted by the theorem.

4. Smoothness and Spectral Accuracy

Program 7

```
% p7.m - accuracy of periodic spectral differentiation

% Compute derivatives for various values of N:
  Nmax = 50; E = zeros(3,Nmax/2-2);
  for N = 6:2:Nmax;
    h = 2*pi/N; x = h*(1:N)';
    column = [0 .5*(-1).^(1:N-1).*cot((1:N-1)*h/2)]';
    D = toeplitz(column,column([1 N:-1:2]));
    v = abs(sin(x)).^3;                     % 3rd deriv in BV
    vprime = 3*sin(x).*cos(x).*abs(sin(x));
    E(1,N/2-2) = norm(D*v-vprime,inf);
    v = exp(-sin(x/2).^(-2));               % C-infinity
    vprime = .5*v.*sin(x)./sin(x/2).^4;
    E(2,N/2-2) = norm(D*v-vprime,inf);
    v = 1./(1+sin(x/2).^2);                 % analytic in a strip
    vprime = -sin(x/2).*cos(x/2).*v.^2;
    E(3,N/2-2) = norm(D*v-vprime,inf);
    v = sin(10*x); vprime = 10*cos(10*x);   % band-limited
    E(4,N/2-2) = norm(D*v-vprime,inf);
  end

% Plot results:
  titles = {'|sin(x)|^3','exp(-sin^{-2}(x/2))',...
      '1/(1+sin^2(x/2))','sin(10x)'}; clf
  for iplot = 1:4
    subplot(2,2,iplot)
    semilogy(6:2:Nmax,E(iplot,:),'.','markersize',12)
    line(6:2:Nmax,E(iplot,:))
    axis([0 Nmax 1e-16 1e3]), grid on
    set(gca,'xtick',0:10:Nmax,'ytick',(10).^(-15:5:0))
    xlabel N, ylabel error, title(titles(iplot))
  end
```

The reader is entitled to be somewhat confused at this point about the many details of Fourier transforms, semidiscrete Fourier transforms, discretizations, inverses, and aliases we have discussed. To see the relationships among these ideas, take a look at the "master plan" presented in Figure 4.2 for the case of an infinite grid. Wander about this diagram and acquaint yourself with every alleyway!

As mentioned earlier, our developments for $\mathbb{R}$ carry over with few changes

Output 7

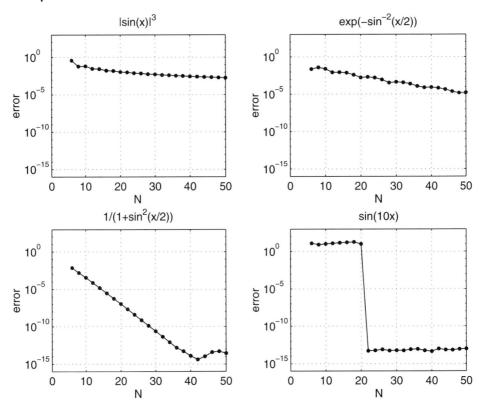

Output 7: *Error as a function of N in the spectral derivatives of four periodic functions. The smoother the function, the faster the convergence.*

to $[-\pi, \pi]$, but as this book is primarily practical, not theoretical, we will not give details. Results on convergence of spectral methods can be found in sources including [CHQZ88], [ReWe99], and [Tad86], of which the last comes closest to following the pattern presented here.

We close this chapter with an example that illustrates spectral accuracy in action. Consider the problem of finding values of λ such that

$$-u'' + x^2 u = \lambda u, \qquad x \in \mathbb{R}, \tag{4.6}$$

for some $u \neq 0$. This is the problem of a quantum harmonic oscillator, whose exact solution is well known [BeOr78]. The eigenvalues are $\lambda = 1, 3, 5, \ldots,$ and the eigenfunctions u are Hermite polynomials multiplied by decreasing exponentials, $e^{-x^2/2} H_n(x)$ (times an arbitrary constant). Since these solutions decay rapidly, for practical computations we can truncate the infinite spatial domain to the periodic interval $[-L, L]$, provided L is sufficiently large. We

4. Smoothness and Spectral Accuracy

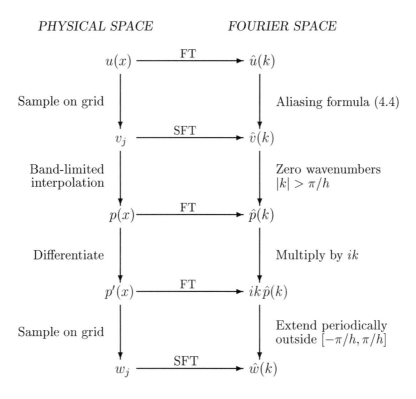

Fig. 4.2. *The master plan of Fourier spectral methods. To get from v_j to w_j, we can stay in physical space or we can cross over to Fourier space and back again. This diagram refers to the spectral method on an infinite grid, but a similar diagram can be constructed for other methods. FT denotes Fourier transform and SFT denotes semidiscrete Fourier transform.*

set up a uniform grid $x_1, \ldots, x_N$ extending across $[-L, L]$, let v be the vector of approximations to u at the grid points, and approximate (4.6) by the matrix equation

$$(-D_N^{(2)} + S)v = \lambda v,$$

where $D_N^{(2)}$ is the second-order periodic spectral differentiation matrix of (3.12) rescaled to $[-L, L]$ instead of $[-\pi, \pi]$ and S is the diagonal matrix

$$S = \mathrm{diag}(x_1^2, \ldots, x_N^2).$$

To approximate the eigenvalues of (4.6) we find the eigenvalues of the matrix $-D_N^{(2)} + S$.

This approximation is constructed in Program 8. Output 8 reveals that the first 4 eigenvalues come out correct to 13 digits on a grid of just 36 points.

Program 8

```
% p8.m - eigenvalues of harmonic oscillator -u"+x^2 u on R

  format long, format compact
  L = 8;                             % domain is [-L L], periodic
  for N = 6:6:36
    h = 2*pi/N; x = h*(1:N); x = L*(x-pi)/pi;
    column = [-pi^2/(3*h^2)-1/6 ...
        -.5*(-1).^(1:N-1)./sin(h*(1:N-1)/2).^2];
    D2 = (pi/L)^2*toeplitz(column);  % 2nd-order differentiation
    eigenvalues = sort(eig(-D2 + diag(x.^2)));
    N, eigenvalues(1:4)
  end
```

Output 8

```
N =  6                              N = 12
     0.46147291699547                   0.97813728129859
     7.49413462105052                   3.17160532064718
     7.72091605300656                   4.45593529116679
    28.83248377834015                   8.92452905811993

N = 18                              N = 24
     0.99997000149932                   0.99999999762904
     3.00064406679582                   3.00000009841085
     4.99259532440770                   4.99999796527330
     7.03957189798150                   7.00002499815654

N = 30                              N = 36
     0.99999999999993                   0.99999999999996
     3.00000000000075                   3.00000000000003
     4.99999999997560                   4.99999999999997
     7.00000000050861                   6.99999999999999
```

Output 8: *Spectrally accurate computed eigenvalues of the harmonic oscillator, with added shading of unconverged digits.*

4. Smoothness and Spectral Accuracy

Summary of This Chapter. Smooth functions have rapidly decaying Fourier transforms, which implies that the aliasing errors introduced by discretization are small. This is why spectral methods are so accurate for smooth functions. In particular, for a function with p derivatives, the νth spectral derivative typically has accuracy $O(h^{p-\nu})$, and for an analytic function, geometric convergence is the rule.

Exercises

4.1. Show that Theorem 3 follows from Theorems 1 and 2.

4.2. Show that Theorem 4 follows from Theorem 3.

4.3. (a) Determine the Fourier transform of $u(x) = (1+x^2)^{-1}$. (Use a complex contour integral if you know how; otherwise, copy the result from (4.3).) (b) Determine $\hat{v}(k)$, where v is the discretization of u on the grid $h\mathbb{Z}$. (*Hint.* Calculating $\hat{v}(k)$ from the definition (2.3) is very difficult.) (c) How fast does $\hat{v}(k)$ approach $\hat{u}(k)$ as $h \to 0$? (d) Does this result match the predictions of Theorem 3?

4.4. Modify Program 7 so that you can verify that the data in the first curve of Output 7 match the prediction of Theorem 4(a). Verify also that the third and fourth curves match the predictions of parts (c) and (d).

4.5. The second curve of Output 7, on the other hand, seems puzzling—we appear to have geometric convergence, yet the function is not analytic. Figure out what is going on. Is the convergence not truly geometric? Or is it geometric for some reason subtler than that which underlies Theorem 4(c), and if so, what is the reason?

4.6. Write a program to investigate the accuracy of Program 8 as a function of L and N/L. On a single plot with a log scale, the program should superimpose twelve curves of $|\lambda_{\text{computed}} - \lambda_{\text{exact}}|$ vs. N/L corresponding to $L = 3, 5, 7$, the lowest four eigenvalues λ, and $N = 4, 6, 8, \ldots, 60$. How large must L and N/L be for the four eigenvalues to be computed to ten-digit accuracy? For sufficiently large L, what is the shape of the convergence curve as a function of N/L? How does this match the results of this chapter and the smoothness of the eigenfunctions being discretized?

4.7. Consider (4.6) with x^2 changed to x^4. What happens to the eigenvalues? Calculate the first 20 of them to 10-digit accuracy, providing good evidence that you have achieved this, and plot the results.

4.8. Derive the Fourier transform of (4.6), and discuss how it relates to (4.6) itself. What does this imply about the functions $\{e^{-x^2/2}H_n(x)\}$?

5. Polynomial Interpolation and Clustered Grids

Of course, not all problems can be treated as periodic. We now begin to consider how to construct spectral methods for bounded, nonperiodic domains.

Suppose that we wish to work on $[-1, 1]$ with nonperiodic functions. One approach would be to pretend that the functions were periodic and use trigonometric (that is, Fourier) interpolation in equispaced points. This is what we did in Program 8. It is a method that works fine for problems like that one whose solutions are exponentially close to zero (or a constant) near the boundaries. In general, however, this approach sacrifices the accuracy advantages of spectral methods. A smooth function

becomes nonsmooth in general when periodically extended:

With a Fourier spectral method, the contamination caused by these discontinuities will be global, destroying the spectral accuracy—the Gibbs phenomenon visible in Output 3 (p. 14). The error in the interpolant will be

$O(1)$, the error in the first derivative will be $O(N)$, and so on. These errors will remain significant even if extra steps are taken to make the functions under study periodic. To achieve good accuracy by a method of that kind it would be necessary to enforce continuity not just of function values but also of several derivatives (see Theorem 4(a), p. 34), a process neither elegant nor efficient.

Instead, it is customary to replace trigonometric polynomials by algebraic polynomials, $p(x) = a_0 + a_1 x + \cdots + a_N x^N$. The first idea we might have is to use polynomial interpolation in equispaced points. Now this, as it turns out, is catastrophically bad in general. A problem known as the *Runge phenomenon* is encountered that is more extreme than the Gibbs phenomenon. When smooth functions are interpolated by polynomials in $N+1$ equally spaced points, the approximations not only fail to converge in general as $N \to \infty$, but they get worse at a rate that may be as great as 2^N. If one were to differentiate such interpolants to compute spectral derivatives, the results would be in error by a similar factor. We shall illustrate this phenomenon in a moment.

The right idea is polynomial interpolation in unevenly spaced points. Various different sets of points are effective, but they all share a common property. Asymptotically as $N \to \infty$, the points are distributed with the density (per unit length)

$$\text{density} \sim \frac{N}{\pi\sqrt{1-x^2}}. \tag{5.1}$$

In particular this implies that the average spacing between points is $O(N^{-2})$ for $x \approx \pm 1$ and $O(N^{-1})$ in the interior, with the average spacing between adjacent points near $x = 0$ asymptotic to π/N. Spectral methods sometimes use points not distributed like this, but in such cases, the interpolants are generally not polynomials but some other functions, such as polynomials stretched by a $\sin^{-1}$ change of variables [For96, KoTa93].

In most of this book we shall use the simplest example of a set of points that satisfy (5.1), the so-called *Chebyshev points*,

$$x_j = \cos(j\pi/N), \qquad j = 0, 1, \ldots, N. \tag{5.2}$$

Geometrically, we can visualize these points as the projections on $[-1, 1]$ of equispaced points on the upper half of the unit circle, as sketched in Figure 5.1. Fuller names for $\{x_j\}$ include *Chebyshev–Lobatto points* and *Gauss–Chebyshev–Lobatto points* (alluding to their role in certain quadrature formulas) and *Chebyshev extreme points* (since they are the extreme points in $[-1, 1]$ of the Chebyshev polynomial $T_N(x)$), but for simplicity, in this book we just call them Chebyshev points.

The effect of using these clustered points on the accuracy of the polynomial interpolant is dramatic. Program 9 interpolates $u(x) = 1/(1 + 16x^2)$ by

5. Polynomial Interpolation and Clustered Grids

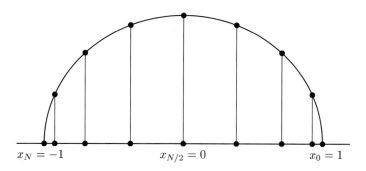

Fig. 5.1. *Chebyshev points are the projections onto the x-axis of equally spaced points on the unit circle. Note that they are numbered from right to left.*

polynomials in both equispaced and Chebyshev points. In Output 9 we see that the former works very badly and the latter very well.

We could stop here and take it as given that for spectral methods based on algebraic polynomials, one must use irregular grids such as (5.2) that have the asymptotic spacing (5.1). However, this fact is so fundamental to the subject of spectral methods—and so interesting!—that we want to explain it. The remainder of this chapter attempts to do this by appealing to the beautiful subject of potential theory.

Suppose we have a monic polynomial p of degree N. We can write it as

$$p(z) = \prod_{k=1}^{N}(z - z_k),$$

where $\{z_k\}$ are the roots, counted with multiplicity, which might be complex. From this formula we have

$$|p(z)| = \prod_{k=1}^{N}|z - z_k|,$$

and therefore

$$\log|p(z)| = \sum_{k=1}^{N}\log|z - z_k|.$$

Now consider

$$\phi_N(z) = N^{-1}\sum_{k=1}^{N}\log|z - z_k|. \qquad (5.3)$$

This function is harmonic in the complex plane (that is, it satisfies Laplace's equation) except at $\{z_k\}$. We can interpret it as an electrostatic potential:

Program 9

```
% p9.m - polynomial interpolation in equispaced and Chebyshev pts
  N = 16;
  xx = -1.01:.005:1.01; clf
  for i = 1:2
    if i==1, s = 'equispaced points'; x = -1 + 2*(0:N)/N; end
    if i==2, s = 'Chebyshev points';   x = cos(pi*(0:N)/N); end
    subplot(2,2,i)
    u = 1./(1+16*x.^2);
    uu = 1./(1+16*xx.^2);
    p = polyfit(x,u,N);             % interpolation
    pp = polyval(p,xx);             % evaluation of interpolant
    plot(x,u,'.','markersize',13)
    line(xx,pp)
    axis([-1.1 1.1 -1 1.5]), title(s)
    error = norm(uu-pp,inf);
    text(-.5,-.5,['max error = ' num2str(error)])
  end
```

Output 9

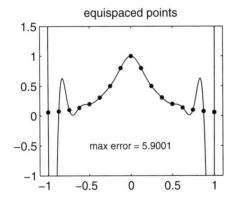
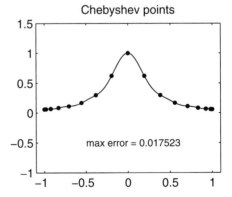

Output 9: *Degree N interpolation of $u(x) = 1/(1+16x^2)$ in $N+1$ equispaced and Chebyshev points for $N = 16$. With increasing N, the errors increase exponentially in the equispaced case—the Runge phenomenon—whereas in the Chebyshev case they decrease exponentially.*

5. Polynomial Interpolation and Clustered Grids

$\phi_N(z)$ is the potential at z due to charges at $\{z_k\}$,
each with potential $N^{-1}\log|z-z_k|$.

By construction, there is a correspondence between the size of $p(z)$ and the value of $\phi_N(z)$,

$$|p(z)| = e^{N\phi_N(z)}. \tag{5.4}$$

From this we can begin to see how the Runge phenomenon is related to potential theory. If $\phi_N(z)$ is approximately constant for $z \in [-1,1]$, then $p(z)$ is approximately constant there too. If $\phi_N(z)$ varies along $[-1,1]$, on the other hand, the effect on $|p(z)|$ will be variations that grow exponentially with N.

In this framework it is natural to take the limit $N \to \infty$ and think in terms of points $\{x_j\}$ distributed in $[-1,1]$ according to a density function $\rho(x)$ with $\int_{-1}^{1} \rho(x)dx = 1$. Such a function gives the number of grid points in an interval $[a,b]$ by the integral

$$N\int_a^b \rho(x)\,dx.$$

For finite N, ρ must be the sum of Dirac delta functions of amplitude N^{-1}, but in the limit, we take it to be smooth. For equispaced points the limit is

$$\rho(x) = \tfrac{1}{2}, \quad x \in [-1,1] \quad \text{(uniform density)}. \tag{5.5}$$

The corresponding potential ϕ is given by the integral

$$\phi(z) = \int_{-1}^{1} \rho(x)\log|z-x|\,dx. \tag{5.6}$$

From this, with a little work, it can be deduced that the potential for equispaced points in the limit $N \to \infty$ is

$$\phi(z) = -1 + \tfrac{1}{2}\mathrm{Re}\big((z+1)\log(z+1) - (z-1)\log(z-1)\big), \tag{5.7}$$

where $\mathrm{Re}(\cdot)$ denotes the real part. Note the particular values $\phi(0) = -1$ and $\phi(\pm 1) = -1 + \log 2$. From these values and (5.4) we conclude that if a polynomial p has roots equally spaced in $[-1,1]$, then it will take values about 2^N times larger near $x = \pm 1$ than near $x = 0$:

$$|p(x)| \simeq e^{N\phi(x)} = \begin{cases} (2/e)^N & \text{near } x = \pm 1, \\ (1/e)^N & \text{near } x = 0. \end{cases}$$

By contrast, consider the continuous distribution corresponding to (5.1):

$$\rho(x) = \frac{1}{\pi\sqrt{1-x^2}}, \quad x \in [-1,1] \quad \text{(Chebyshev density)}. \tag{5.8}$$

With a little work this gives the potential

$$\phi(z) = \log \frac{|z - \sqrt{z^2-1}|}{2}. \tag{5.9}$$

This formula has a simple interpretation: the level curves of $\phi(z)$ are the ellipses with foci ± 1, and the value of $\phi(z)$ along any such ellipse is the logarithm of half the sum of the semimajor and semiminor axes. In particular, the degenerate ellipse $[-1,1]$ is a level curve where $\phi(z)$ takes the constant value $-\log 2$ (Exercise 5.5). We conclude that if a monic polynomial p has N roots spaced according to the Chebyshev distribution in $[-1,1]$, then it will oscillate between values of comparable size on the order of 2^{-N} throughout $[-1,1]$:

$$|p(x)| \simeq e^{N\phi(x)} = 2^{-N}, \qquad x \in [-1,1].$$

Program 10

```
% p10.m - polynomials and corresponding equipotential curves
  N = 16; clf
  for i = 1:2
    if i==1, s = 'equispaced points'; x = -1 + 2*(0:N)/N; end
    if i==2, s = 'Chebyshev points';   x = cos(pi*(0:N)/N); end
    p = poly(x);

    % Plot p(x) over [-1,1]:
    xx = -1:.005:1; pp = polyval(p,xx);
    subplot(2,2,2*i-1)
    plot(x,0*x,'.','markersize',13), hold on
    plot(xx,pp), grid on
    set(gca,'xtick',-1:.5:1), title(s)

    % Plot equipotential curves:
    subplot(2,2,2*i)
    plot(real(x),imag(x),'.','markersize',13), hold on
    axis([-1.4 1.4 -1.12 1.12])
    xgrid = -1.4:.02:1.4; ygrid = -1.12:.02:1.12;
    [xx,yy] = meshgrid(xgrid,ygrid); zz = xx+1i*yy;
    pp = polyval(p,zz); levels = 10.^(-4:0);
    contour(xx,yy,abs(pp),levels), title(s), colormap([0 0 0])
  end
```

5. Polynomial Interpolation and Clustered Grids

Output 10

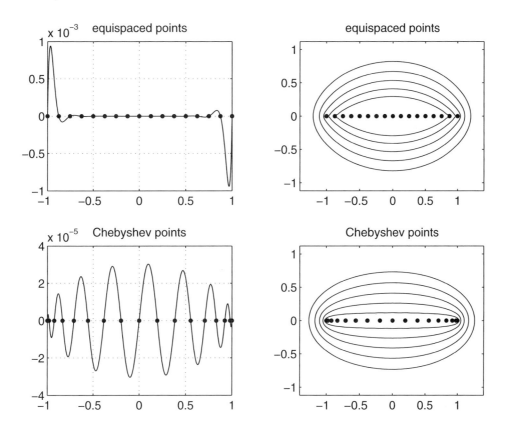

Output 10: *On the left, the degree* 17 *monic polynomials with equispaced and Chebyshev roots. On the right, some level curves of the corresponding potentials in the complex plane. Chebyshev points are good because they generate a potential for which* $[-1, 1]$ *is approximately a level curve.*

Program 10 illustrates these relationships. The first plot of Output 10 shows the degree 17 monic polynomial defined by equispaced roots on $[-1,1]$, revealing large swings near the boundary. The plot to the right shows the corresponding potential, and we see that $[-1,1]$ is not close to an equipotential curve. The bottom pair presents analogous results for the Chebyshev case. Now p oscillates on the same scale throughout $[-1,1]$, and $[-1,1]$ is close to an equipotential curve. (It would exactly equioscillate, if we had defined Chebyshev points as the zeros rather than the extrema of Chebyshev polynomials.)

Though we will not give proofs, much more can be concluded from this kind of analysis:

Theorem 5 Accuracy of polynomial interpolation.

Given a function u and a sequence of sets of interpolation points $\{x_j\}_N$, $N = 1, 2, \ldots$, that converge to a density function ρ as $n \to \infty$ with corresponding potential ϕ given by (5.6), define

$$\phi_{[-1,1]} = \sup_{x \in [-1,1]} \phi(x).$$

For each N construct the polynomial p_N of degree $\leq N$ that interpolates u at the points $\{x_j\}_N$. If there exists a constant $\phi_u > \phi_{[-1,1]}$ such that u is analytic throughout the closed region

$$\{z \in \mathbb{C} : \phi(z) \leq \phi_u\},$$

then there exists a constant $C > 0$ such that for all $x \in [-1, 1]$ and all N,

$$|u(x) - p_N(x)| \leq C\, e^{-N(\phi_u - \phi_{[-1,1]})}.$$

The same estimate holds, though with a new constant C (still independent of x and N), for the difference of the νth derivatives, $u^{(\nu)} - p_N^{(\nu)}$, for any $\nu \geq 1$.

In a word, polynomial interpolants and spectral methods converge geometrically (in the absence of rounding errors), provided u is analytic in a neighborhood of the region bounded by the smallest equipotential curve that contains $[-1, 1]$. Conversely, for equally spaced points we must expect divergence for functions that are not analytic throughout the "football" (American football, that is!) of the upper right plot of Output 10 that just passes through ± 1. The function f of Program 9 has poles at $\pm i/4$, inside the football, which explains the divergence of equispaced interpolation for that function (Exercise 5.1).

Theorem 5 is stated in considerable generality, and it is worthwhile recording the special form it takes in the situation we most care about, namely spectral differentiation in Chebyshev points. Here, the level curves of ϕ are ellipses, and we get the following result, sharper variants of which can be found in [ReWe99] and [Tad86].

Theorem 6 Accuracy of Chebyshev spectral differentiation.

Suppose u is analytic on and inside the ellipse with foci ± 1 on which the Chebyshev potential takes the value ϕ_f, that is, the ellipse whose semimajor and semiminor axis lengths sum to $K = e^{\phi_f + \log 2}$. Let w be the νth Chebyshev spectral derivative of u ($\nu \geq 1$). Then

$$|w_j - u^{(\nu)}(x_j)| = O(e^{-N(\phi_f + \log 2)}) = O(K^{-N})$$

as $N \to \infty$.

5. Polynomial Interpolation and Clustered Grids

We say that the *asymptotic convergence factor* for the spectral differentiation process is at least as small as K^{-1}:

$$\limsup_{N\to\infty} |w_j - u^{(\nu)}(x_j)|^{1/N} \leq K^{-1}.$$

This is not the end of the story. Where does the Chebyshev charge distribution "really" come from? One answer comes by noting that if a potential ϕ is constant on $[-1, 1]$, then $\phi'(z) = 0$ on $[-1, 1]$. If we think of $\phi'(x)$, the gradient of a potential, as a force that will be exerted on a unit charge, we conclude that

The Chebyshev density function ρ of (5.8) *is an equilibrium, minimal-energy distribution for a unit charge distributed continuously on* $[-1, 1]$.

For finite N, a monic polynomial p of minimax size in $[-1, 1]$ will not generally have roots exactly at equilibrium points in $[-1, 1]$, but as $N \to \infty$, it can be proved that the roots must converge to a density function $\rho(x)$ with this distribution (5.8). This continuum limit is normally discussed by defining ϕ to be the *Green's function* for $[-1, 1]$, the unique real function that takes a constant value on $[-1, 1]$, is harmonic outside it, and is asymptotic to $\log |z|$ as $|z| \to \infty$. For more on this beautiful mathematical subject, see [EmTr99], [Hil62], and [Tsu59].

Summary of This Chapter. Grid points for polynomial spectral methods should lie approximately in a minimal-energy configuration associated with inverse linear repulsion between points. On $[-1, 1]$, this means clustering near $x = \pm 1$ according to the Chebyshev distribution (5.1). For a function u analytic on $[-1, 1]$, the corresponding spectral derivatives converge geometrically, with an asymptotic convergence factor determined by the size of the largest ellipse about $[-1, 1]$ in which u is analytic.

Exercises

5.1. Modify Program 9 to compute and plot the maximum error over $[-1, 1]$ for equispaced and Chebyshev interpolation on a log scale as a function of N. What asymptotic divergence and convergence constants do you observe for these two cases? (Confine your attention to small enough values of N that rounding errors are not dominant.) Now, based on the potential theory of this chapter, determine exactly what these geometric constants should be. How closely do your numerical experiments match the theoretical answers?

5.2. Modify Program 9 to measure the error in equispaced interpolation of $u(z) = 1/(1 + 16z^2)$ not just on $[-1, 1]$ but on a grid in the rectangular complex domain $-1.2 < \mathrm{Re}\, z, \mathrm{Im}\, z < 1.2$. The absolute values of the errors should then be visualized by a contour plot, and the region where their error is $< 10^{-2}$ should be shaded. The poles of $u(z)$ should also be marked. How does the picture look for $N = 10, 20, 30$?

5.3. Let $E_N = \inf_p \|p(x) - e^x\|_\infty$, where $\|f\|_\infty = \sup_{x \in [-1,1]} |f(x)|$, denote the error in degree N minimax polynomial approximation to e^x on $[-1, 1]$. (a) One candidate approximation $p(x)$ would be the Taylor series truncated at degree N. From this approximation, derive the bound $E_N < ((N+2)/(N+1))/(N+1)!$ for $N \geq 0$. (b) In fact, the truncated Taylor series falls short of optimal by a factor of about 2^N, for it is known (see equation (6.75) of [Mei67]) that as $N \to \infty$, $E_N \sim 2^{-N}/(N+1)!$. Modify Program 9 to produce a plot showing this asymptotic formula, the upper bound of (a), the error when $p(x)$ is obtained from interpolation in Chebyshev points, and the same for equispaced points, all as a function of N for $N = 1, 2, 3, \ldots, 12$. Comment on the results.

5.4. Derive (5.7).

5.5. Derive (5.9), and show that $\phi(z)$ has the constant value $-\log 2$ on $[-1, 1]$.

5.6. Potentials and Green's functions associated with connected sets in the complex plane can be obtained by conformal mapping. For example, the Chebyshev points are images of roots of unity on the unit circle under a conformal map of the exterior of the unit disk to the exterior of $[-1, 1]$. Determine this conformal map and use it to derive (5.9).

5.7. In continuation of the last exercise, for polynomial interpolation on a polygonal set P in the complex plane, good sets of interpolation points can be obtained by a Schwarz–Christoffel conformal map of the exterior of the unit disk to the exterior of P. Download Driscoll's MATLAB Schwarz–Christoffel Toolbox [Dri96] and get it to work on your machine. Use it to produce a plot of 20 good interpolation points on an equilateral triangle and on another polygonal domain P of your choosing. Pick a point z_0 lying a little bit outside P and use your points to interpolate $u(z) = (z - z_0)^{-1}$. How big is the maximum error on the boundary of P? (By the maximum modulus principle, this is the same as the error in the interior.) How does this compare with the error if you interpolate in equally spaced points along the boundary of P?

6. Chebyshev Differentiation Matrices

In the last chapter we discussed why grid points must cluster at boundaries for spectral methods based on polynomials. In particular, we introduced the Chebyshev points,

$$x_j = \cos(j\pi/N), \qquad j = 0, 1, \ldots, N, \qquad (6.1)$$

which cluster as required. In this chapter we shall use these points to construct Chebyshev differentiation matrices and apply these matrices to differentiate a few functions. The same set of points will continue to be the basis of many of our computations throughout the rest of the book.

Our scheme is as follows. Given a grid function v defined on the Chebyshev points, we obtain a discrete derivative w in two steps:

- *Let p be the unique polynomial of degree $\leq N$ with $p(x_j) = v_j$, $0 \leq j \leq N$.*
- *Set $w_j = p'(x_j)$.*

This operation is linear, so it can be represented by multiplication by an $(N+1) \times (N+1)$ matrix, which we shall denote by D_N:

$$w = D_N v.$$

Here N is an arbitrary positive integer, even or odd. The restriction to even N in this book (p. 18) applies to Fourier, not Chebyshev spectral methods.

To get a feel for the interpolation process, we take a look at $N = 1$ and $N = 2$ before proceeding to the general case.

Consider first $N = 1$. The interpolation points are $x_0 = 1$ and $x_1 = -1$, and the interpolating polynomial through data v_0 and v_1, written in Lagrange form, is

$$p(x) = \tfrac{1}{2}(1+x)v_0 + \tfrac{1}{2}(1-x)v_1.$$

Taking the derivative gives

$$p'(x) = \tfrac{1}{2}v_0 - \tfrac{1}{2}v_1.$$

This formula implies that D_1 is the 2×2 matrix whose first column contains constant entries $1/2$ and whose second column contains constant entries $-1/2$:

$$D_1 = \begin{pmatrix} \tfrac{1}{2} & -\tfrac{1}{2} \\ \tfrac{1}{2} & -\tfrac{1}{2} \end{pmatrix}.$$

Now consider $N = 2$. The interpolation points are $x_0 = 1$, $x_1 = 0$, and $x_2 = -1$, and the interpolant is the quadratic

$$p(x) = \tfrac{1}{2}x(1+x)v_0 + (1+x)(1-x)v_1 + \tfrac{1}{2}x(x-1)v_2.$$

The derivative is now a linear polynomial,

$$p'(x) = (x + \tfrac{1}{2})v_0 - 2xv_1 + (x - \tfrac{1}{2})v_2.$$

The differentiation matrix D_2 is the 3×3 matrix whose jth column is obtained by sampling the jth term of this expression at $x = 1, 0$, and -1:

$$D_2 = \begin{pmatrix} \tfrac{3}{2} & -2 & \tfrac{1}{2} \\ \tfrac{1}{2} & 0 & -\tfrac{1}{2} \\ -\tfrac{1}{2} & 2 & -\tfrac{3}{2} \end{pmatrix}. \tag{6.2}$$

It is no coincidence that the middle row of this matrix contains the coefficients for a centered three-point finite difference approximation to a derivative, and the other rows contain the coefficients for one-sided approximations such as the one that drives the second-order Adams–Bashforth formula for the numerical solution of ODEs [For88]. The rows of higher order spectral differentiation matrices can also be viewed as vectors of coefficients of finite difference formulas, but these will be based on uneven grids and thus no longer familiar from standard applications.

We now give formulas for the entries of D_N for arbitrary N. These were first published perhaps in [GHO84] and are derived in Exercises 6.1 and 6.2. Analogous formulas for general sets $\{x_j\}$ rather than just Chebyshev points are stated in Exercise 6.1.

6. Chebyshev Differentiation Matrices

Theorem 7 Chebyshev differentiation matrix.

For each $N \geq 1$, let the rows and columns of the $(N+1) \times (N+1)$ Chebyshev spectral differentiation matrix D_N be indexed from 0 to N. The entries of this matrix are

$$(D_N)_{00} = \frac{2N^2+1}{6}, \qquad (D_N)_{NN} = -\frac{2N^2+1}{6}, \qquad (6.3)$$

$$(D_N)_{jj} = \frac{-x_j}{2(1-x_j^2)}, \qquad j = 1, \ldots, N-1, \qquad (6.4)$$

$$(D_N)_{ij} = \frac{c_i}{c_j} \frac{(-1)^{i+j}}{(x_i - x_j)}, \qquad i \neq j, \quad i,j = 0, \ldots, N, \qquad (6.5)$$

where

$$c_i = \begin{cases} 2, & i = 0 \text{ or } N, \\ 1, & \text{otherwise.} \end{cases}$$

A picture makes the pattern clearer:

$$D_N = \begin{bmatrix} \dfrac{2N^2+1}{6} & 2\dfrac{(-1)^j}{1-x_j} & \dfrac{1}{2}(-1)^N \\[1em] -\dfrac{1}{2}\dfrac{(-1)^i}{1-x_i} & \dfrac{-x_j}{2(1-x_j^2)} \quad \dfrac{(-1)^{i+j}}{x_i-x_j} & \dfrac{1}{2}\dfrac{(-1)^{N+i}}{1+x_i} \\[1em] -\dfrac{1}{2}(-1)^N & -2\dfrac{(-1)^{N+j}}{1+x_j} & -\dfrac{2N^2+1}{6} \end{bmatrix}$$

The jth column of D_N contains the derivative of the degree N polynomial interpolant $p_j(x)$ to the delta function supported at x_j, sampled at the grid

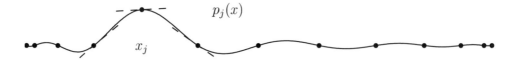

Fig. 6.1. *Degree 12 polynomial interpolant $p(x)$ to the delta function supported at x_8 on the 13-point Chebyshev grid with $N = 12$. The slopes indicated by the dashed lines, from right to left, are the entries $(D_{12})_{7,8}$, $(D_{12})_{8,8}$, and $(D_{12})_{9,8}$ of the 13×13 spectral differentiation matrix D_{12}.*

points $\{x_i\}$. Three such sampled values are suggested by the dashed lines in Figure 6.1.

Throughout this text, we take advantage of MATLAB's high-level commands for such operations as polynomial interpolation, matrix inversion, and FFT. For clarity of exposition, as explained in the "Note on the MATLAB Programs" at the beginning of the book, our style is to make our programs short and self-contained. However, there will be one major exception to this rule, one MATLAB function that we will define and then call repeatedly whenever we need Chebyshev grids and differentiation matrices. The function is called cheb, and it returns a vector x and a matrix D.

cheb.m

```
% CHEB   compute D = differentiation matrix, x = Chebyshev grid

  function [D,x] = cheb(N)
  if N==0, D=0; x=1; return, end
  x = cos(pi*(0:N)/N)';
  c = [2; ones(N-1,1); 2].*(-1).^(0:N)';
  X = repmat(x,1,N+1);
  dX = X-X';
  D  = (c*(1./c)')./(dX+(eye(N+1)));    % off-diagonal entries
  D  = D - diag(sum(D'));                % diagonal entries
```

Note that this program does not compute D_N exactly by formulas (6.3)–(6.5). It utilizes (6.5) for the off-diagonal entries but then obtains the diagonal

6. Chebyshev Differentiation Matrices

entries (6.3)–(6.4) from the identity

$$(D_N)_{ii} = -\sum_{\substack{j=0 \\ j \neq i}}^{N} (D_N)_{ij}. \qquad (6.6)$$

This is marginally simpler to program, and it produces a matrix with better stability properties in the presence of rounding errors [BaBe00, BCM94]. Equation (6.6) can be derived by noting that the interpolant to $(1,1,\ldots,1)^T$ is the constant function $p(x) = 1$, and since $p'(x) = 0$ for all x, D_N must map $(1,1,\ldots,1)^T$ to the zero vector.

Here are the first five Chebyshev differentiation matrices as computed by **cheb**. Note that they are dense, with little apparent structure apart from the antisymmetry condition $(D_N)_{ij} = -(D_N)_{N-i,N-j}$.

```
>> cheb(1)
    0.5000   -0.5000
    0.5000   -0.5000

>> cheb(2)
    1.5000   -2.0000    0.5000
    0.5000   -0.0000   -0.5000
   -0.5000    2.0000   -1.5000

>> cheb(3)
    3.1667   -4.0000    1.3333   -0.5000
    1.0000   -0.3333   -1.0000    0.3333
   -0.3333    1.0000    0.3333   -1.0000
    0.5000   -1.3333    4.0000   -3.1667

>> cheb(4)
    5.5000   -6.8284    2.0000   -1.1716    0.5000
    1.7071   -0.7071   -1.4142    0.7071   -0.2929
   -0.5000    1.4142   -0.0000   -1.4142    0.5000
    0.2929   -0.7071    1.4142    0.7071   -1.7071
   -0.5000    1.1716   -2.0000    6.8284   -5.5000

>> cheb(5)
    8.5000  -10.4721    2.8944   -1.5279    1.1056   -0.5000
    2.6180   -1.1708   -2.0000    0.8944   -0.6180    0.2764
   -0.7236    2.0000   -0.1708   -1.6180    0.8944   -0.3820
    0.3820   -0.8944    1.6180    0.1708   -2.0000    0.7236
   -0.2764    0.6180   -0.8944    2.0000    1.1708   -2.6180
    0.5000   -1.1056    1.5279   -2.8944   10.4721   -8.5000
```

Program 11 illustrates how D_N can be used to differentiate the smooth, nonperiodic function $u(x) = e^x \sin(5x)$ on grids with $N = 10$ and $N = 20$. The output shows a graph of $u(x)$ alongside a plot of the error in $u'(x)$. With $N = 20$, we get nine-digit accuracy.

Program 11

```
% p11.m - Chebyshev differentation of a smooth function
  xx = -1:.01:1; uu = exp(xx).*sin(5*xx); clf
  for N = [10 20]
    [D,x] = cheb(N); u = exp(x).*sin(5*x);
      subplot('position',[.15 .66-.4*(N==20) .31 .28])
      plot(x,u,'.','markersize',14), grid on
      line(xx,uu)
      title(['u(x),  N=' int2str(N)])
    error = D*u - exp(x).*(sin(5*x)+5*cos(5*x));
      subplot('position',[.55 .66-.4*(N==20) .31 .28])
      plot(x,error,'.','markersize',14), grid on
      line(x,error)
      title(['  error in u''(x),  N=' int2str(N)])
  end
```

Output 11

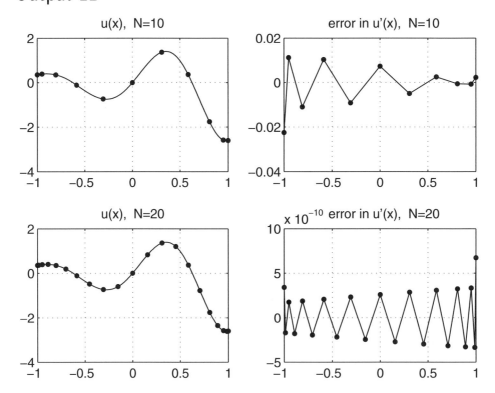

Output 11: *Chebyshev differentiation of $u(x) = e^x \sin(5x)$. Note the vertical scales.*

6. Chebyshev Differentiation Matrices

Program 12

```
% p12.m - accuracy of Chebyshev spectral differentiation
%         (compare p7.m)

% Compute derivatives for various values of N:
  Nmax = 50; E = zeros(3,Nmax);
  for N = 1:Nmax;
    [D,x] = cheb(N);
    v = abs(x).^3; vprime = 3*x.*abs(x);      % 3rd deriv in BV
    E(1,N) = norm(D*v-vprime,inf);
    v = exp(-x.^(-2)); vprime = 2.*v./x.^3;   % C-infinity
    E(2,N) = norm(D*v-vprime,inf);
    v = 1./(1+x.^2); vprime = -2*x.*v.^2;     % analytic in [-1,1]
    E(3,N) = norm(D*v-vprime,inf);
    v = x.^10; vprime = 10*x.^9;              % polynomial
    E(4,N) = norm(D*v-vprime,inf);
  end

% Plot results:
  titles = {'|x^3|','exp(-x^{-2})','1/(1+x^2)','x^{10}'}; clf
  for iplot = 1:4
    subplot(2,2,iplot)
    semilogy(1:Nmax,E(iplot,:),'.','markersize',12)
    line(1:Nmax,E(iplot,:))
    axis([0 Nmax 1e-16 1e3]), grid on
    set(gca,'xtick',0:10:Nmax,'ytick',(10).^(-15:5:0))
    xlabel N, ylabel error, title(titles(iplot))
  end
```

Program 12, the Chebyshev analogue of Program 7, illustrates spectral accuracy more systematically. Four functions are spectrally differentiated: $|x^3|$, $\exp(-x^{-2})$, $1/(1+x^2)$, and x^{10}. The first has a third derivative of bounded variation, the second is smooth but not analytic, the third is analytic in a neighborhood of $[-1,1]$, and the fourth is a polynomial, the analogue for Chebyshev spectral methods of a band-limited function for Fourier spectral methods.

Summary of This Chapter. The entries of the Chebyshev differentiation matrix D_N can be computed by explicit formulas, which can be conveniently collected in an eight-line MATLAB function. More general explicit formulas can be used to construct the differentiation matrix for an arbitrarily prescribed set of distinct points $\{x_j\}$.

Output 12

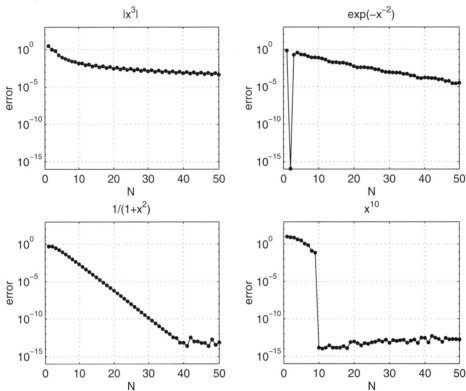

Output 12: *Accuracy of the Chebyshev spectral derivative for four functions of increasing smoothness. Compare Output 7 (p. 36).*

Exercises

6.1. If $x_0, x_1, \ldots, x_N \in \mathbb{R}$ are distinct, then the *cardinal function* $p_j(x)$ defined by

$$p_j(x) = \frac{1}{a_j} \prod_{\substack{k=0 \\ k \neq j}}^{N} (x - x_k), \qquad a_j = \prod_{\substack{k=0 \\ k \neq j}}^{N} (x_j - x_k) \tag{6.7}$$

is the unique polynomial interpolant of degree N to the values 1 at x_j and 0 at x_k, $k \neq j$. Take the logarithm and differentiate to obtain

$$p_j'(x) = p_j(x) \sum_{\substack{k=0 \\ k \neq j}}^{N} (x - x_k)^{-1},$$

and from this derive the formulas

6. Chebyshev Differentiation Matrices

$$D_{ij} = \frac{1}{a_j} \prod_{\substack{k=0 \\ k \neq i,j}}^{N} (x_i - x_k) = \frac{a_i}{a_j(x_i - x_j)} \qquad (i \neq j) \qquad (6.8)$$

and

$$D_{jj} = \sum_{\substack{k=0 \\ k \neq j}}^{N} (x_j - x_k)^{-1} \qquad (6.9)$$

for the entries of the $N \times N$ differentiation matrix associated with the points $\{x_j\}$. (See also Exercise 12.2.)

6.2. Derive Theorem 7 from (6.8) and (6.9).

6.3. Suppose $1 = x_0 > x_1 > \cdots > x_N = -1$ lie in the minimal-energy configuration in $[-1, 1]$ in the sense discussed on p. 49. Show that except in the corners, the diagonal entries of the corresponding differentiation matrix D are zero.

6.4. It was mentioned on p. 55 that Chebyshev differentiation matrices have the symmetry property $(D_N)_{ij} = -(D_N)_{N-i,N-j}$. (a) Explain where this condition comes from. (b) Derive the analogous symmetry condition for $(D_N)^2$. (c) Taking N to be odd, so that the dimension of D_N is even, explain how $(D_N)^2$ could be constructed from just half the entries of D_N. For large N, how does the floating point operation count for this process compare with that for straightforward squaring of D_N?

6.5. Modify cheb so that it computes the diagonal entries of D_N by the explicit formulas (6.3)–(6.4) rather than by (6.6). Confirm that your code produces the same results except for rounding errors. Then see if you can find numerical evidence that it is less stable numerically than cheb.

6.6. The second panel of Output 12 shows a sudden dip for $N = 2$. Show that in fact, $E(2, 2) = 0$ (apart from rounding errors).

6.7. Theorem 6 makes a prediction about the geometric rate of convergence in the third panel of Output 12. Exactly what is this prediction? How well does it match the observed rate of convergence?

6.8. Let D_N be the usual Chebyshev differentiation matrix. Show that the power $(D_N)^{N+1}$ is identically equal to zero. Now try it on the computer for $N = 5$ and 20 and report the computed 2-norms $\|(D_5)^6\|_2$ and $\|(D_{20})^{21}\|_2$. Discuss.

7. Boundary Value Problems

We have defined the Chebyshev differentiation matrix D_N and put together a MATLAB program, `cheb`, to compute it. In this chapter we illustrate how such matrices can be used to solve some boundary value problems arising in ordinary and partial differential equations.

As our first example, consider the linear ODE boundary value problem

$$u_{xx} = e^{4x}, \quad -1 < x < 1, \quad u(\pm 1) = 0. \tag{7.1}$$

This is a Poisson equation, with solution $u(x) = [e^{4x} - x\sinh(4) - \cosh(4)]/16$. We use the PDE notation u_{xx} instead of u'' because we shall soon increase the number of dimensions.

To solve the problem numerically, we can compute the second derivative via D_N^2, the square of D_N. The first thing to note is that D_N^2 can be evaluated either by squaring D_N, which costs $O(N^3)$ floating point operations, or by explicit formulas [GoLu83a, Pey86] or recurrences [WeRe00, Wel97], which cost $O(N^2)$ floating point operations. There are real advantages to the latter approaches, but in this book, for simplicity, we just square D_N.

The other half of the problem is the imposition of the boundary conditions $u(\pm 1) = 0$. For simple problems like (7.1) with homogeneous Dirichlet boundary conditions, we can proceed as follows. We take the interior Chebyshev points $x_1, \ldots, x_{N-1}$ as our computational grid, with $v = (v_1, \ldots, v_{N-1})^T$ as the corresponding vector of unknowns. Spectral differentiation is then carried out like this:

- Let $p(x)$ be the unique polynomial of degree $\leq N$ with $p(\pm 1) = 0$ and $p(x_j) = v_j$, $1 \leq j \leq N - 1$.

- *Set $w_j = p''(x_j)$, $1 \le j \le N-1$.*

This is not the only means of imposing boundary conditions in spectral methods. We shall consider alternatives in Chapter 13, where among other examples, Programs 32 and 33 (pp. 136 and 138) solve (7.1) again with inhomogeneous Dirichlet and homogeneous Neumann boundary conditions, respectively.

Now D_N^2 is an $(N+1) \times (N+1)$ matrix that maps a vector $(v_0, \ldots, v_N)^T$ to a vector $(w_0, \ldots, w_N)^T$. The procedure just described amounts to a decision that we wish to:

- *Fix v_0 and v_N at zero.*
- *Ignore w_0 and w_N.*

This implies that the first and last columns of D_N^2 have no effect (since multiplied by zero) and the first and last rows have no effect either (since ignored):

In other words, to solve our one-dimensional Poisson problem by a Chebyshev spectral method, we can make use of the $(N-1) \times (N-1)$ matrix $\widetilde{D}_N^2$ obtained by stripping D_N^2 of its first and last rows and columns. In MATLAB notation:

$$\widetilde{D}_N^2 = D_N^2(1\colon N-1,\, 1\colon N-1).$$

In an actual MATLAB program, since indices start at 1 instead of 0, this will become D2 = D2(2:N,2:N) or D2 = D2(2:end-1,2:end-1).

With $\widetilde{D}_N^2$ in hand, the numerical solution of (7.1) becomes a matter of solving a linear system of equations:

$$\widetilde{D}_N^2 v = f.$$

Program 13 carries out this process. We should draw attention to a feature of this program that appears here for the first time in the book and will reappear in a number of our later programs. Although the algorithm calculates the vector $(v_1, \ldots, v_{N-1})^T$ of approximations to u at the grid points, as always with

7. Boundary Value Problems

spectral methods, we really have more information about the numerical solution than just point values. Implicitly we are dealing with a polynomial interpolant $p(x)$, and in MATLAB this can be calculated conveniently (though not very quickly or stably) by a command of the form `polyval(polyfit(...))`. Program 13 uses this trick to evaluate $p(x)$ on a fine grid, both for plotting and for measuring the error, which proves to be on the order of 10^{-10}. Exercise 7.1 investigates the more stable method for constructing $p(x)$ known as barycentric interpolation. For practical plotting purposes with spectral methods, much simpler local interpolants are usually adequate; see, e.g., the use of `interp2( ... ,'cubic')` in Program 16 (p. 70).

What if the equation is nonlinear? For example, suppose we change (7.1) to

$$u_{xx} = e^u, \qquad -1 < x < 1, \quad u(\pm 1) = 0. \qquad (7.2)$$

Because of the nonlinearity, it is no longer enough simply to invert the second-order differentiation matrix $\widetilde{D}_N^2$. Instead, we can solve the problem iteratively. We choose an initial guess, such as the vector of zeros, and then iterate by repeatedly solving the system of equations

$$\widetilde{D}_N^2 v_{\text{new}} = \exp(v_{\text{old}}),$$

where $\exp(v)$ is the column vector defined componentwise by $(\exp(v))_j = e^{v_j}$. Program 14 implements this iteration with a crude stopping criterion, and convergence occurs in 29 steps.

To convince ourselves that we have obtained the correct solution, we can modify Program 14 to print results for various N. Here is such a table:

N	no. its.	u(0)
2	34	-0.35173371124920
4	29	-0.36844814823915
6	29	-0.36805450387666
8	29	-0.36805614384219
10	29	-0.36805602345302
12	29	-0.36805602451189
14	29	-0.36805602444069
16	29	-0.36805602444149
18	30	-0.36805602444143
20	29	-0.36805602444143

Evidently $u(0)$ is accurate to 12 or 13 digits, even with $N = 16$. The convergence of this iteration is analyzed in Exercise 7.3.

As a third application of the modified second-order differentiation matrix $\widetilde{D}_N^2$, consider the eigenvalue boundary value problem

$$u_{xx} = \lambda u, \qquad -1 < x < 1, \quad u(\pm 1) = 0. \qquad (7.3)$$

Program 13

```
% p13.m - solve linear BVP u_xx = exp(4x), u(-1)=u(1)=0
  N = 16;
  [D,x] = cheb(N);
  D2 = D^2;
  D2 = D2(2:N,2:N);                    % boundary conditions
  f = exp(4*x(2:N));
  u = D2\f;                            % Poisson eq. solved here
  u = [0;u;0];
  clf, subplot('position',[.1 .4 .8 .5])
  plot(x,u,'.','markersize',16)
  xx = -1:.01:1;
  uu = polyval(polyfit(x,u,N),xx);     % interpolate grid data
  line(xx,uu)
  grid on
  exact = ( exp(4*xx) - sinh(4)*xx - cosh(4) )/16;
  title(['max err = ' num2str(norm(uu-exact,inf))],'fontsize',12)
```

Output 13

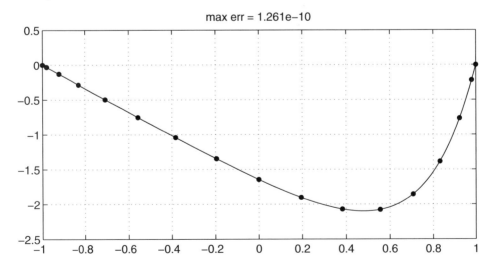

Output 13: *Solution of the linear boundary value problem* (7.1).

7. Boundary Value Problems

Program 14

```
% p14.m - solve nonlinear BVP u_xx = exp(u), u(-1)=u(1)=0
%         (compare p13.m)
  N = 16;
  [D,x] = cheb(N); D2 = D^2; D2 = D2(2:N,2:N);
  u = zeros(N-1,1);
  change = 1; it = 0;
  while change > 1e-15                    % fixed-point iteration
    unew = D2\exp(u);
    change = norm(unew-u,inf);
    u = unew; it = it+1;
  end
  u = [0;u;0];
  clf, subplot('position',[.1 .4 .8 .5])
  plot(x,u,'.','markersize',16)
  xx = -1:.01:1;
  uu = polyval(polyfit(x,u,N),xx);
  line(xx,uu), grid on
  title(sprintf('no. steps = %d    u(0) =%18.14f',it,u(N/2+1)))
```

Output 14

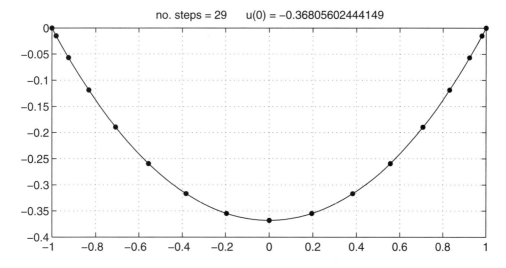

Output 14: *Solution of the nonlinear boundary value problem (7.2).*

Program 15

```
% p15.m - solve eigenvalue BVP u_xx = lambda*u, u(-1)=u(1)=0

  N = 36; [D,x] = cheb(N); D2 = D^2; D2 = D2(2:N,2:N);
  [V,Lam] = eig(D2); lam = diag(Lam);
  [foo,ii] = sort(-lam);          % sort eigenvalues and -vectors
  lam = lam(ii); V = V(:,ii); clf
  for j = 5:5:30                  % plot 6 eigenvectors
    u = [0;V(:,j);0]; subplot(7,1,j/5)
    plot(x,u,'.','markersize',12), grid on
    xx = -1:.01:1; uu = polyval(polyfit(x,u,N),xx);
    line(xx,uu), axis off
    text(-.4,.5,sprintf('eig %d =%20.13f*4/pi^2',j,lam(j)*4/pi^2))
    text(.7,.5,sprintf('%4.1f  ppw', 4*N/(pi*j)))
  end
```

Output 15

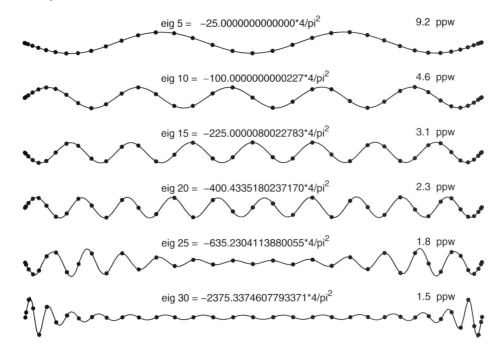

Output 15: *Eigenvalues and eigenmodes of $\widetilde{D}_N^2$ and the number of grid points per wavelength (ppw) at the center of the grid.*

7. Boundary Value Problems

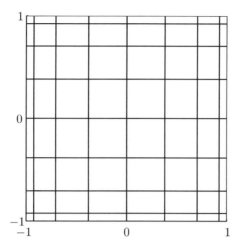

Fig. 7.1. *A tensor product grid.*

The eigenvalues of this problem are $\lambda = -\pi^2 n^2/4$, $n = 1, 2, \ldots$, with corresponding eigenfunctions $\sin(n\pi(x+1)/2)$. Program 15 calculates the eigenvalues and eigenvectors of $\widetilde{D}_N^2$ for $N = 36$ by MATLAB's built-in matrix eigenvalue routine `eig`. The numbers and plots of Output 15 reveal a great deal about the accuracy of spectral methods. Eigenvalues 5, 10, and 15 are obtained to many digits of accuracy, and eigenvalue 20 is still pretty good. Eigenvalue 25 is accurate to only one digit, however, and eigenvalue 30 is wrong by a factor of 3. The crucial quantity that explains this behavior is the number of points per wavelength ("ppw") in the central, coarsest part of the grid near $x = 0$. With at least two points per wavelength, the grid is fine enough everywhere to resolve the wave. With less than two points per wavelength, the wave cannot be resolved, and eigenvectors are obtained that are meaningless as approximations to the original problem.

We now consider how to extend these methods to boundary value problems in several space dimensions. To be specific, here is a two-dimensional Poisson problem:

$$u_{xx} + u_{yy} = 10\sin(8x(y-1)), \quad -1 < x, y < 1, \quad u = 0 \text{ on the boundary.} \tag{7.4}$$

(The right-hand side has been chosen to make an interesting picture.) For such a problem we naturally set up a grid based on Chebyshev points independently in each direction, called a *tensor product grid* (Figure 7.1). Note that whereas in one dimension, a Chebyshev grid is $2/\pi$ times as dense in the middle as an equally spaced grid, in d dimensions this figure becomes $(2/\pi)^d$. Thus the great majority of grid points lie near the boundary. Sometimes this is wasteful, and techniques have been devised to reduce the waste [For96,

KaSh99, KoTa93]. At other times, when boundary layers or other fine details appear near boundaries, the extra resolution there may be useful.

The easiest way to solve a problem on a tensor product spectral grid is to use tensor products in linear algebra, also known as *Kronecker products*. The Kronecker product of two matrices A and B is denoted by $A \otimes B$ and is computed in MATLAB by the command `kron(A,B)`. If A and B are of dimensions $p \times q$ and $r \times s$, respectively, then $A \otimes B$ is the matrix of dimension $pr \times qs$ with $p \times q$ block form, where the i,j block is $a_{ij}B$. For example,

$$\begin{pmatrix} 1 & 2 \\ 3 & 4 \end{pmatrix} \otimes \begin{pmatrix} a & b \\ c & d \end{pmatrix} = \left(\begin{array}{cc|cc} a & b & 2a & 2b \\ c & d & 2c & 2d \\ \hline 3a & 3b & 4a & 4b \\ 3c & 3d & 4c & 4d \end{array} \right).$$

To explain how Kronecker products can be used for spectral methods, let us consider the case $N = 4$. Suppose we number the internal nodes in the obvious, "lexicographic" ordering:

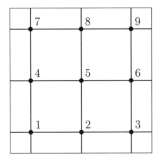

Also suppose that we have data $(v_1, v_2, \ldots, v_9)^T$ at these grid points. We wish to approximate the Laplacian by differentiating spectrally in the x and y directions independently. Now the 3×3 differentiation matrix with $N = 4$ in one dimension is given by `D = cheb(4); D2 = D^2; D2 = D2(2:4,2:4)`:

$$\widetilde{D}_4^2 = \begin{pmatrix} -14 & 6 & -2 \\ 4 & -6 & 4 \\ -2 & 6 & -14 \end{pmatrix}.$$

If I denotes the 3×3 identity, then the second derivative with respect to x

will accordingly be computed by the matrix kron(I,D2):

$$
I \otimes \widetilde{D}_N^2 = \left(\begin{array}{ccc|ccc|ccc}
-14 & 6 & -2 & & & & & & \\
4 & -6 & 4 & & & & & & \\
-2 & 6 & -14 & & & & & & \\
\hline
 & & & -14 & 6 & -2 & & & \\
 & & & 4 & -6 & 4 & & & \\
 & & & -2 & 6 & -14 & & & \\
\hline
 & & & & & & -14 & 6 & -2 \\
 & & & & & & 4 & -6 & 4 \\
 & & & & & & -2 & 6 & -14
\end{array} \right).
$$

The second derivative with respect to y will be computed by kron(D2,I):

$$
\widetilde{D}_N^2 \otimes I = \left(\begin{array}{ccc|ccc|ccc}
-14 & & & 6 & & & -2 & & \\
 & -14 & & & 6 & & & -2 & \\
 & & -14 & & & 6 & & & -2 \\
\hline
4 & & & -6 & & & 4 & & \\
 & 4 & & & -6 & & & 4 & \\
 & & 4 & & & -6 & & & 4 \\
\hline
-2 & & & 6 & & & -14 & & \\
 & -2 & & & 6 & & & -14 & \\
 & & -2 & & & 6 & & & -14
\end{array} \right).
$$

Our discrete Laplacian is now the *Kronecker sum* [HoJo91]

$$L_N = I \otimes \widetilde{D}_N^2 + \widetilde{D}_N^2 \otimes I. \tag{7.5}$$

This matrix, though not dense, is not as sparse as one typically gets with finite differences or finite elements. Fortunately, thanks to spectral accuracy, we may hope to obtain satisfactory results with dimensions in the hundreds rather than the thousands or tens of thousands.

Program 16 solves the Poisson problem (7.4) numerically with $N = 24$. The program produces two plots, which we label Output 16a and Output 16b. The first shows the locations of the 23,805 nonzero entries in the 529×529 matrix L_{24}. The second plots the solution and prints the value $u(x,y)$ for $x = y = 2^{-1/2}$, which is convenient because this is one of the grid points whenever N is divisible by 4. The program also notes the time taken to perform the solution of the linear system of equations: on my Sparc Ultra 5 workstation in MATLAB version 6.0, 1.2 seconds.

A variation of the Poisson equation is the *Helmholtz equation*,

$$u_{xx} + u_{yy} + k^2 u = f(x,y), \quad -1 < x, y < 1, \quad u = 0 \text{ on the boundary}, \tag{7.6}$$

Program 16

```
% p16.m - Poisson eq. on [-1,1]x[-1,1] with u=0 on boundary

% Set up grids and tensor product Laplacian and solve for u:
  N = 24; [D,x] = cheb(N); y = x;
  [xx,yy] = meshgrid(x(2:N),y(2:N));
  xx = xx(:); yy = yy(:);          % stretch 2D grids to 1D vectors
  f = 10*sin(8*xx.*(yy-1));
  D2 = D^2; D2 = D2(2:N,2:N); I = eye(N-1);
  L = kron(I,D2) + kron(D2,I);                        % Laplacian
  figure(1), clf, spy(L), drawnow
  tic, u = L\f; toc          % solve problem and watch the clock

% Reshape long 1D results onto 2D grid:
  uu = zeros(N+1,N+1); uu(2:N,2:N) = reshape(u,N-1,N-1);
  [xx,yy] = meshgrid(x,y);
  value = uu(N/4+1,N/4+1);

% Interpolate to finer grid and plot:
  [xxx,yyy] = meshgrid(-1:.04:1,-1:.04:1);
  uuu = interp2(xx,yy,uu,xxx,yyy,'cubic');
  figure(2), clf, mesh(xxx,yyy,uuu), colormap([0 0 0])
  xlabel x, ylabel y, zlabel u
  text(.4,-.3,-.3,sprintf('u(2^{-1/2},2^{-1/2}) = %14.11f',value))
```

where k is a real parameter. This equation arises in the analysis of wave propagation governed by the equation

$$-U_{tt} + U_{xx} + U_{yy} = e^{ikt} f(x,y), \quad -1 < x,y < 1, \quad U = 0 \text{ on the boundary} \tag{7.7}$$

after separation of variables to get $U(x,y,t) = e^{ikt} u(x,y)$. Program 17 is a minor modification of Program 16 to solve such a problem for the particular choices

$$k = 9, \qquad f(x,y) = \exp(-10\left[(y-1)^2 + (x - \tfrac{1}{2})^2\right]). \tag{7.8}$$

The solution appears as a mesh plot in Output 17a and as a contour plot in Output 17b. It is clear that the response generated by this forcing function $f(x,y)$ for this value $k=9$ has approximately the form of a wave with three half-wavelengths in the x direction and five half-wavelengths in the y direction. This is easily explained. Such a wave is an eigenfunction of the homogeneous

7. Boundary Value Problems

Output 16a

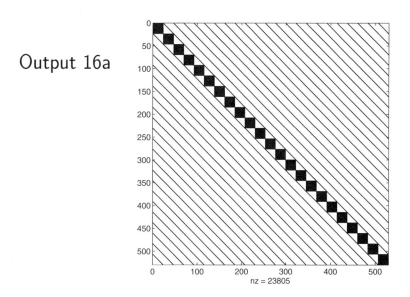

Output 16a: *Sparsity plot of the 529×529 discrete Laplacian* (7.5).

Output 16b

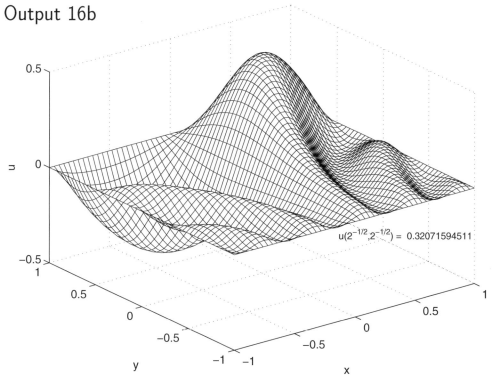

Output 16b: *Solution of the Poisson equation* (7.4). *The result has been interpolated to a finer rectangular grid for plotting. The computed value $u(2^{-1/2}, 2^{-1/2})$ is accurate to nine digits.*

Program 17

```
% p17.m - Helmholtz eq. u_xx + u_yy + (k^2)u = f
%          on [-1,1]x[-1,1]    (compare p16.m)

% Set up spectral grid and tensor product Helmholtz operator:
  N = 24; [D,x] = cheb(N); y = x;
  [xx,yy] = meshgrid(x(2:N),y(2:N));
  xx = xx(:); yy = yy(:);
  f = exp(-10*((yy-1).^2+(xx-.5).^2));
  D2 = D^2; D2 = D2(2:N,2:N); I = eye(N-1);
  k = 9;
  L = kron(I,D2) + kron(D2,I) + k^2*eye((N-1)^2);

% Solve for u, reshape to 2D grid, and plot:
  u = L\f;
  uu = zeros(N+1,N+1); uu(2:N,2:N) = reshape(u,N-1,N-1);
  [xx,yy] = meshgrid(x,y);
  [xxx,yyy] = meshgrid(-1:.0333:1,-1:.0333:1);
  uuu = interp2(xx,yy,uu,xxx,yyy,'cubic');
  figure(1), clf, mesh(xxx,yyy,uuu), colormap([0 0 0])
  xlabel x, ylabel y, zlabel u
  text(.2,1,.022,sprintf('u(0,0) = %13.11f',uu(N/2+1,N/2+1)))
  figure(2), clf, contour(xxx,yyy,uuu)
  colormap([0 0 0]), axis square
```

Helmholtz problem (i.e., $f(x,y) = 0$) with eigenvalue

$$k = \tfrac{1}{2}\sqrt{3^2 + 5^2} \approx 9.1592.$$

Our choice $k = 9$ gives near-resonance with this (3,5) mode.

Summary of This Chapter. Homogeneous Dirichlet boundary conditions for spectral collocation methods can be implemented by simply deleting the first and/or last rows and columns of a spectral differentiation matrix. Problems in two space dimensions can be formulated in terms of Kronecker products, and for moderate-sized grids, they can be solved that way on the computer. Nonlinear problems can be solved by iteration.

7. Boundary Value Problems

Output 17a

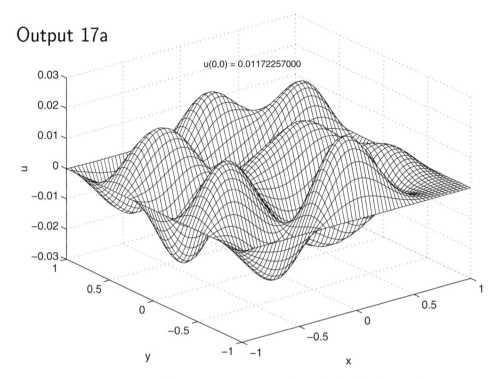

Output 17a: *Solution of the Helmholtz problem* (7.6), (7.8). *The computed value $u(0,0)$ is accurate to nine digits.*

Output 17b

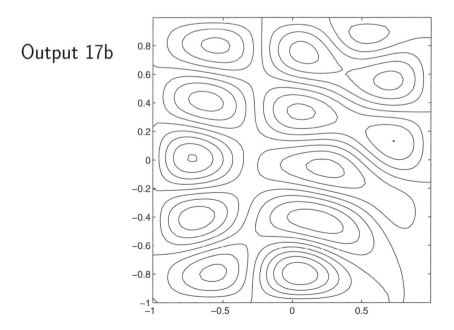

Output 17b: *Same result represented as contour plot.*

Exercises

7.1. Modify Program 13 so that instead of `polyval` and `polyfit`, it uses the more stable formula of *barycentric interpolation* [Hen82]:

$$p(x) = \sum_{j=0}^{N} \frac{a_j^{-1} u_j}{x - x_j} \bigg/ \sum_{j=0}^{N} \frac{a_j^{-1}}{x - x_j}, \qquad (7.9)$$

where $\{a_j\}$ are defined by (6.7). Experiment with various interpolation problems (such as that of Exercise 5.1) and find evidence of the enhanced stability of this method.

7.2. Solve the boundary value problem $u_{xx} + 4u_x + e^x u = \sin(8x)$ numerically on $[-1, 1]$ with boundary conditions $u(\pm 1) = 0$. To 10 digits of accuracy, what is $u(0)$?

7.3. In the iteration of Program 14, each step is observed to reduce the error norm by a factor of about 0.2943. This explains why 30 steps are enough to reduce the error to 10^{-14}. Add one or two lines to the code to compute the eigenvalues of an appropriate matrix to show where the number 0.2943 comes from.

7.4. Devise an alternative to Program 14 based on Newton iteration rather than fixed-point iteration, and make it work. Do you observe quadratic convergence?

7.5. A curious feature of Program 15 is that, although the problem is self-adjoint, the matrix that approximates it is not symmetric. This is typical of spectral collocation (but not Galerkin) methods. Many things can be said about how much it does or does not matter [CaGo96, McRo00], but let us consider just one: the cost in linear algebra. Perform experiments in MATLAB to estimate how much slower nonsymmetric real eigenvalue/eigenvector calculations are than symmetric ones for dense $N \times N$ matrices for values of N such as 100, 200, 300. Look up the algorithms in a book such as [Dem97] or [GoVa96] to see how your experiments match theoretical predictions.

7.6. Show how, by adding just two characters to Program 16, one can make the program solve the linear system of equations by sparse rather than dense methods of numerical linear algebra. This particular sparsity structure is not readily exploited, however. Provide evidence on this matter by comparing timings for the dense and sparse variants of the code with $N = 24$ and 32.

7.7. As explained in the text, the solution of Output 17 has the form it does because of near-resonance with the (5,3) eigenvalue $k \approx 9.1592$. Run the same program to produce contour plots for each of the integers $k = 1, 2, 3, \ldots, 20$. In each case, judge from the figure what mode (i, j), if any, seems to be principally excited, and produce a table showing how closely k matches the associated eigenvalue $(\pi/2)\sqrt{i^2 + j^2}$.

8. Chebyshev Series and the FFT

In this chapter we will see how Chebyshev spectral methods can be implemented by the FFT, which provides a crucial speedup for some calculations. Equally important will be the mathematical idea that underlies this technique: the equivalence of

$$\begin{aligned}
&\text{Chebyshev series} &&\text{in } x \in [-1, 1], \\
&\text{Fourier series} &&\text{in } \theta \in \mathbb{R}, \\
&\text{Laurent series} &&\text{in } z \text{ on the unit circle.}
\end{aligned}$$

The basis of our development is summarized in Figure 8.1. Let z be a complex number on the unit circle: $|z| = 1$. Let θ be the argument of z, a real number that is determined up to multiples of 2π. Let $x = \text{Re}\, z = \cos\theta$. For each $x \in [-1, 1]$, there are two complex conjugate values of z, and we have

$$x = \text{Re}\, z = \tfrac{1}{2}(z + z^{-1}) = \cos\theta \in [-1, 1]. \tag{8.1}$$

The nth *Chebyshev polynomial*, denoted T_n, is defined by

$$T_n(x) = \text{Re}\, z^n = \tfrac{1}{2}(z^n + z^{-n}) = \cos n\theta. \tag{8.2}$$

From this formula, it is not obvious that $T_n(x)$ is a polynomial in x. The cases $n = 0, 1, 2,$ and 3 make the point clear:

$$\text{Re}\, z^0 = 1 \quad \Rightarrow \quad T_0(x) = 1,$$

$$\text{Re}\, z^1 = \tfrac{1}{2}(z + z^{-1}) \quad \Rightarrow \quad T_1(x) = x,$$

$$\text{Re}\, z^2 = \tfrac{1}{2}(z^2 + z^{-2}) = \tfrac{1}{2}(z^1 + z^{-1})^2 - 1 \quad \Rightarrow \quad T_2(x) = 2x^2 - 1,$$

$$\text{Re}\, z^3 = \tfrac{1}{2}(z^3 + z^{-3}) = \tfrac{1}{2}(z^1 + z^{-1})^3 - \tfrac{3}{2}(z + z^{-1}) \quad \Rightarrow \quad T_3(x) = 4x^3 - 3x.$$

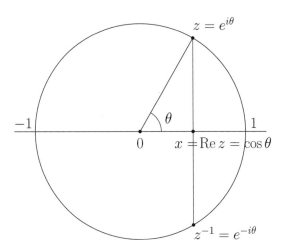

Fig. 8.1. *Relationships between x, z, and θ.*

In general,
$$T_{n+1}(x) = \tfrac{1}{2}(z^{n+1} + z^{-n-1}) = \tfrac{1}{2}(z^n + z^{-n})(z + z^{-1}) - \tfrac{1}{2}(z^{n-1} + z^{1-n}),$$
which amounts to the recurrence relation
$$T_{n+1}(x) = 2xT_n(x) - T_{n-1}(x). \tag{8.3}$$

By induction, we deduce that T_n is a polynomial of degree exactly n for each $n \geq 0$, with leading coefficient 2^{n-1} for each $n \geq 1$. Figure 8.2 gives a geometric interpretation.

Since T_n is of exact degree n for each n, any degree N polynomial can be written uniquely as a linear combination of Chebyshev polynomials,
$$p(x) = \sum_{n=0}^{N} a_n T_n(x), \qquad x \in [-1, 1]. \tag{8.4}$$

Corresponding to this is a degree N *Laurent polynomial* in z and z^{-1} that is *self-reciprocal*, which means that z^n and z^{-n} have equal coefficients:
$$\mathsf{p}(z) = \tfrac{1}{2}\sum_{n=0}^{N} a_n(z^n + z^{-n}), \qquad |z| = 1. \tag{8.5}$$

Also corresponding to these is a degree N 2π-periodic trigonometric polynomial that is even, that is, such that $P(\theta) = P(-\theta)$:
$$P(\theta) = \sum_{n=0}^{N} a_n \cos n\theta, \qquad \theta \in \mathbb{R}. \tag{8.6}$$

8. Chebyshev Series and the FFT

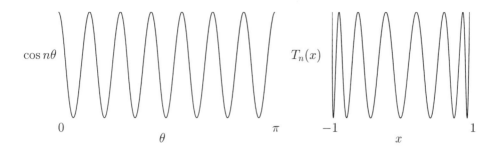

Fig. 8.2. *The Chebyshev polynomial T_n can be interpreted as a sine wave "wrapped around a cylinder and viewed from the side."*

The functions (8.4)–(8.6) are equivalent in the sense that $p(x) = \mathsf{p}(z) = P(\theta)$ when x, z, and θ are related by (8.1). Note that we have introduced different fonts to distinguish the x, z, and θ domains. Similarly, from an arbitrary function $f(x)$ defined for $x \in [-1, 1]$, we can form a self-reciprocal function $\mathsf{f}(z)$ defined on the unit circle and a periodic function $F(\theta)$ defined on $\mathbb{R}$:

$$\mathsf{f}(z) = f\left(\frac{z + z^{-1}}{2}\right), \qquad F(\theta) = f(\cos\theta).$$

For spectral collocation methods, we mainly deal with (8.4)–(8.6) as interpolants of function f, f, and F. The interpolation points are as follows:

$$\theta_j = j\pi/N,$$

$$z_j = e^{i\theta_j},$$

$$x_j = \cos\theta_j = \operatorname{Re} z_j,$$

with $0 \leq j \leq N$. We have the equivalences:

$P(\theta)$ interpolates $F(\theta)$ (even and 2π-periodic) in the equispaced points $\{\theta_j\}$

$\Updownarrow$

$\mathsf{p}(z)$ interpolates $\mathsf{f}(z)$ (self-reciprocal) in the roots of unity $\{z_j\}$

$\Updownarrow$

$p(x)$ interpolates $f(x)$ (arbitrary) in the Chebyshev points $\{x_j\}$.

We are now prepared to describe an FFT algorithm for Chebyshev spectral differentiation. The key point is that the polynomial interpolant q of f can be differentiated by finding a trigonometric polynomial interpolant Q of F, differentiating in Fourier space, and transforming back to the x variable. Once

Chebyshev spectral differentiation via FFT

- Given data $v_0, \ldots, v_N$ at Chebyshev points $x_0 = 1, \ldots, x_N = -1$, extend this data to a vector V of length $2N$ with $V_{2N-j} = v_j$, $j = 1, 2, \ldots, N-1$.

- Using the FFT, calculate
$$\hat{V}_k = \frac{\pi}{N} \sum_{j=1}^{2N} e^{-ik\theta_j} V_j, \qquad k = -N+1, \ldots, N.$$

- Define $\hat{W}_k = ik\hat{V}_k$, except $\hat{W}_N = 0$.

- Compute the derivative of the trigonometric interpolant Q on the equispaced grid by the inverse FFT:
$$W_j = \frac{1}{2\pi} \sum_{k=-N+1}^{N} e^{ik\theta_j} \hat{W}_k, \qquad j = 1, \ldots, 2N.$$

- Calculate the derivative of the algebraic polynomial interpolant q on the interior grid points by
$$w_j = -\frac{W_j}{\sqrt{1-x_j^2}}, \qquad j = 1, \ldots, N-1,$$
with the special formulas at the endpoints
$$w_0 = \frac{1}{2\pi} {\sum_{n=0}^{N}}' n^2 \hat{v}_n, \qquad w_N = \frac{1}{2\pi} {\sum_{n=0}^{N}}' (-1)^{n+1} n^2 \hat{v}_n,$$
where the prime indicates that the terms $n = 0, N$ are multiplied by $\frac{1}{2}$.

These formulas can be explained as follows. The trigonometric interpolant of the extended $\{v_j\}$ data is given by evaluating the inverse FFT at arbitrary θ. Using the a_n coefficients we find that
$$P(\theta) = \frac{1}{2\pi} \sum_{k=-N+1}^{N} e^{ik\theta} \hat{V}_k = \sum_{n=0}^{N} a_n \cos n\theta.$$

8. Chebyshev Series and the FFT

The algebraic polynomial interpolant of the $\{v_j\}$ data is $p(x) = P(\theta)$, where $x = \cos\theta$, and the derivative is

$$q'(x) = \frac{Q'(\theta)}{dx/d\theta} = \frac{-\sum_{n=0}^{N} na_n \sin n\theta}{-\sin\theta} = \frac{\sum_{n=0}^{N} na_n \sin n\theta}{\sqrt{1-x^2}}.$$

As for the special formulas for w_0 and w_N, we determine the value of $q'(x)$ at $x = \pm 1$ by l'Hôpital's rule [Str91], which gives

$$q'(1) = \sum_{n=0}^{N} n^2 a_n, \qquad q'(-1) = \sum_{n=0}^{N} (-1)^{n+1} n^2 a_n.$$

It is straightforward to generalize the method for higher derivatives. At the stage of differentiation in Fourier space we multiply by $(ik)^\nu$ to calculate the νth derivative, and if ν is odd, we set $\hat{W}_N = 0$. Secondly, the appropriate factors need to be calculated for converting between derivatives on the equispaced grid and on the Chebyshev grid, i.e., derivatives in the θ and x variables. For example, the second derivatives are related by

$$q''(x) = \frac{-x}{(1-x^2)^{3/2}} Q'(\theta) + \frac{1}{1-x^2} Q''(\theta). \tag{8.7}$$

If W_j and $W_j^{(2)}$ are the first and second derivatives on the equispaced grid, respectively, then the second derivative on the Chebyshev grid is given by

$$w_j^{(2)} = \frac{-x_j}{(1-x_j^2)^{3/2}} W_j + \frac{1}{1-x_j^2} W_j^{(2)}, \qquad 1 \leq j \leq N-1.$$

Again, special formulas are needed for $j = 0$ and N.

On p. 24 it was mentioned that when the complex FFT is applied to differentiate a real periodic function, a factor of 2 in efficiency is lost. In the method we have just described, the situation is worse, for not only is V real (typically), but it is even (always), and together these facts imply that $\hat{V}$ is real and even too (Exercise 2.2). A factor of 4 is now at stake, and the right way to take advantage of this is to use a *discrete cosine transform (DCT)* instead of an FFT. See [BrHe95], [Van92], and Appendix F of [For96] for a discussion of symmetries in Fourier transforms and how to take advantage of them. At the time of this writing, however, although a DCT code is included in MATLAB's Signal Processing Toolbox, there is no DCT in MATLAB itself. In the following program, `chebfft`, we have accordingly chosen to use the general FFT code and accept the loss of efficiency.

chebfft.m

```
% CHEBFFT  Chebyshev differentiation via FFT. Simple, not optimal.
%          If v is complex, delete "real" commands.

  function w = chebfft(v)
  N = length(v)-1; if N==0, w=0; return, end
  x = cos((0:N)'*pi/N);
  ii = 0:N-1;
  v = v(:); V = [v; flipud(v(2:N))];       % transform x -> theta
  U = real(fft(V));
  W = real(ifft(1i*[ii 0 1-N:-1]'.*U));
  w = zeros(N+1,1);
  w(2:N) = -W(2:N)./sqrt(1-x(2:N).^2);     % transform theta -> x
  w(1) = sum(ii'.^2.*U(ii+1))/N + .5*N*U(N+1);
  w(N+1) = sum((-1).^(ii+1)'.*ii'.^2.*U(ii+1))/N + ...
              .5*(-1)^(N+1)*N*U(N+1);
```

Program 18 calls `chebfft` to calculate the Chebyshev derivative of $f(x) = e^x \sin(5x)$ for $N = 10$ and 20 using the FFT. The results are given in Output 18. Compare this with Output 11 (p. 56), which illustrates the same calculation implemented using matrices. The differences are just at the level of rounding errors.

To see the method at work for a PDE, consider the wave equation

$$u_{tt} = u_{xx}, \quad -1 < x < 1, \quad t > 0, \quad u(\pm 1) = 0. \tag{8.8}$$

To solve this equation numerically we use a leap frog formula in t and Chebyshev spectral differentiation in x. To complete the formulation of the numerical method we need to specify two initial conditions. For the PDE, these would typically be conditions on u and u_t. For the finite difference scheme, we need conditions on u at $t = 0$ and at $t = -\Delta t$, the previous time step. Our choice at $t = -\Delta t$ is initial data corresponding to a left-moving Gaussian pulse. Program 19 implements this and should be compared with Program 6 (p. 26). This program, however, runs rather slowly because of the short time step $\Delta t \approx 0.0013$ needed for numerical stability. Time step restrictions are discussed in Chapter 10.

As a second example we consider the wave equation in two space dimensions:

$$u_{tt} = u_{xx} + u_{yy}, \quad -1 < x, y < 1, \quad t > 0, \quad u = 0 \text{ on the boundary}, \tag{8.9}$$

with initial data

$$u(x, y, 0) = e^{-40((x-0.4)^2 + y^2)}, \quad u_t(x, y, 0) = 0.$$

8. Chebyshev Series and the FFT

Program 18

```
% p18.m - Chebyshev differentiation via FFT (compare p11.m)
  xx = -1:.01:1; ff = exp(xx).*sin(5*xx); clf
  for N = [10 20]
    x = cos(pi*(0:N)'/N); f = exp(x).*sin(5*x);
    subplot('position',[.15 .66-.4*(N==20) .31 .28])
    plot(x,f,'.','markersize',14), grid on
    line(xx,ff)
    title(['f(x), N=' int2str(N)])
    error = chebfft(f) - exp(x).*(sin(5*x)+5*cos(5*x));
    subplot('position',[.55 .66-.4*(N==20) .31 .28])
    plot(x,error,'.','markersize',14), grid on
    line(x,error)
    title(['error in f''(x), N=' int2str(N)])
  end
```

Output 18

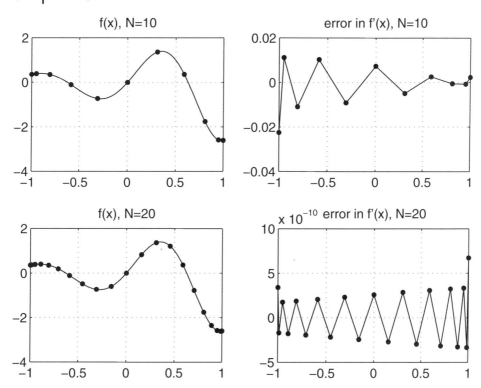

Output 18: *Chebyshev differentiation of $e^x \sin(5x)$ via FFT. Compare Output 11 (p. 56), based on matrices.*

Program 19

```
% p19.m - 2nd-order wave eq. on Chebyshev grid (compare p6.m)

% Time-stepping by leap frog formula:
  N = 80; x = cos(pi*(0:N)/N); dt = 8/N^2;
  v = exp(-200*x.^2); vold = exp(-200*(x-dt).^2);
  tmax = 4; tplot = .075;
  plotgap = round(tplot/dt); dt = tplot/plotgap;
  nplots = round(tmax/tplot);
  plotdata = [v; zeros(nplots,N+1)]; tdata = 0;
  clf, drawnow, h = waitbar(0,'please wait...');
  for i = 1:nplots, waitbar(i/nplots)
    for n = 1:plotgap
      w = chebfft(chebfft(v))'; w(1) = 0; w(N+1) = 0;
      vnew = 2*v - vold + dt^2*w; vold = v; v = vnew;
    end
    plotdata(i+1,:) = v; tdata = [tdata; dt*i*plotgap];
  end

% Plot results:
  clf, drawnow, waterfall(x,tdata,plotdata)
  axis([-1 1 0 tmax -2 2]), view(10,70), grid off
  colormap([0 0 0]), ylabel t, zlabel u, close(h)
```

Output 19

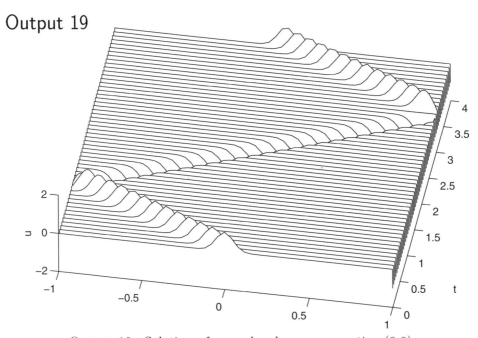

Output 19: *Solution of second-order wave equation* (8.8).

8. Chebyshev Series and the FFT   83

Program 20

```
% p20.m - 2nd-order wave eq. in 2D via FFT (compare p19.m)

% Grid and initial data:
  N = 24; x = cos(pi*(0:N)/N); y = x';
  dt = 6/N^2;
  [xx,yy] = meshgrid(x,y);
  plotgap = round((1/3)/dt); dt = (1/3)/plotgap;
  vv = exp(-40*((xx-.4).^2 + yy.^2));
  vvold = vv;

% Time-stepping by leap frog formula:
  [ay,ax] = meshgrid([.56 .06],[.1 .55]); clf
  for n = 0:3*plotgap
    t = n*dt;
    if rem(n+.5,plotgap)<1     % plots at multiples of t=1/3
      i = n/plotgap+1;
      subplot('position',[ax(i) ay(i) .36 .36])
      [xxx,yyy] = meshgrid(-1:1/16:1,-1:1/16:1);
      vvv = interp2(xx,yy,vv,xxx,yyy,'cubic');
      mesh(xxx,yyy,vvv), axis([-1 1 -1 1 -0.15 1])
      colormap([0 0 0]), title(['t = ' num2str(t)]), drawnow
    end
    uxx = zeros(N+1,N+1); uyy = zeros(N+1,N+1);
    ii = 2:N;
    for i = 2:N                % 2nd derivs wrt x in each row
      v = vv(i,:); V = [v fliplr(v(ii))];
      U = real(fft(V));
      W1 = real(ifft(1i*[0:N-1 0 1-N:-1].*U)); % diff wrt theta
      W2 = real(ifft(-[0:N 1-N:-1].^2.*U));    % diff^2 wrt theta
      uxx(i,ii) = W2(ii)./(1-x(ii).^2) - x(ii).* ...
                  W1(ii)./(1-x(ii).^2).^(3/2);
    end
    for j = 2:N                % 2nd derivs wrt y in each column
      v = vv(:,j); V = [v; flipud(v(ii))];
      U = real(fft(V));
      W1 = real(ifft(1i*[0:N-1 0 1-N:-1]'.*U));% diff wrt theta
      W2 = real(ifft(-[0:N 1-N:-1]'.^2.*U));   % diff^2 wrt theta
      uyy(ii,j) = W2(ii)./(1-y(ii).^2) - y(ii).* ...
                  W1(ii)./(1-y(ii).^2).^(3/2);
    end
    vvnew = 2*vv - vvold + dt^2*(uxx+uyy);
    vvold = vv; vv = vvnew;
  end
```

Output 20a

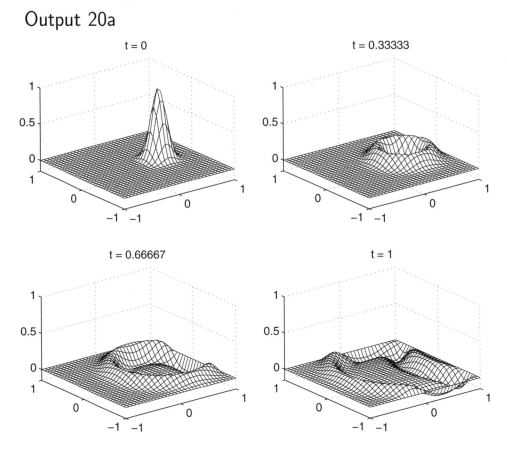

Output 20a: *Solution of the second-order wave equation* (8.9) *on a square.*

Program 20 discretizes this problem with a leap frog formula again for the time discretization and a Chebyshev spectral method on a tensor product grid in x and y.

As in Programs 13–17, the solutions of Output 20a have been interpolated to a finer grid than the computational one for a better display. (Global polynomial interpolation would give spectral accuracy, but here, for simplicity, we use MATLAB's local `interp2` command.) They may accordingly give a misleading impression that the computational grid is regular. In fact, it is a spectral grid, clustered at the boundaries, and to emphasize this point, Output 20b shows the raw data at $t = 1/3$ on the spectral grid. This plot was obtained by replacing `mesh(xxx,yyy,vvv)` in Program 20 by `mesh(xx,yy,vv)`. We emphasize that despite the apparent crudity of this plot, the results at each grid point are spectrally accurate, aside from errors of magnitude $O((\Delta t)^2)$ introduced by time-stepping.

8. Chebyshev Series and the FFT

Output 20b

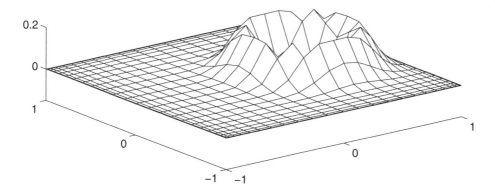

Output 20b: *Same as the second plot of Output 20, but without the interpolation to a finer grid.*

Summary of This Chapter. Chebyshev differentiation can be carried out by the FFT. The underlying idea is that of transplantation from Chebyshev points on $[-1, 1]$ to equally spaced points on the unit circle. Ideally, for real data, a real discrete cosine transform (DCT) should be used, but the general FFT is also applicable with a certain loss of efficiency.

Exercises

8.1. For $n \geq 1$, the leading coefficient of $T_n(x)$ is $c_n = 2^{n-1}$, and this implies $\lim_{n \to \infty} (c_n)^{1/N} = 2$. Derive this limit from the general results of Chapter 5.

8.2. Perform a study of the relative efficiencies of cheb and chebfft as a function of N. Do not count the time taken by cheb to form D_N, just the time taken to multiply D_N by a vector.

8.3. Modify Program 12 (p. 57) to make use of chebfft instead of cheb. The results should be the same as in Output 12, except for rounding errors. Are the effects of rounding errors smaller or larger than before?

8.4. Modify Program 20 to make use of matrices instead of the FFT. Make sure to do this elegantly, using matrix-matrix multiplications rather than explicit loops. You will find that the code gets much shorter, and faster, too. How much faster is it? Does increasing N from 24 to 48 tip the balance?

8.5. Find a way to modify your program of Exercise 8.4, as in Exercise 3.8 but now for a second-order problem, to make use of matrix exponentials rather than time discretization. What effect does this have on the computation time?

8.6. Write a code chebfft2 for second-order differentiation by the FFT, and show by examples that it matches the results obtained by matrices, apart from rounding errors.

8.7. Explain how the entries of the Chebyshev differentiation matrix D_N could be computed by suitable calls to `chebfft` rather than by explicit formulas as in Theorem 7 (p. 53). What is the asymptotic operation count for this method as $N \to \infty$?

8.8. Write a code `chebdct` that computes the same result as `chebfft`, but makes use of the function DCT from MATLAB's Signal Processing Toolbox. (The code will be restricted to real data.) What gain of efficiency do you observe?

8.9. Suppose $p(x) = \sum_{n=0}^{N} a_n T_n(x) = \sum_{n=0}^{N} c_n x^n$, and consider the vectors $a = (a_0, \ldots, a_N)^T$ and $c = (c_0, \ldots, c_N)^T$. Let A be the $(N+1) \times (N+1)$ matrix such that $c = Aa$. Write down A in the case $N = 3$. Now write a short and elegant MATLAB function `cheb2mon` such that the command `A = cheb2mon(N)` constructs this matrix A. Use `cheb2mon` to determine what polynomial $\sum_{n=0}^{5} c_n x^n$ corresponds to $T_0(x) - 2T_1(x) + 3T_2(x) + 2T_3(x) + T_4(x) - T_5(x)$ and what polynomial $\sum_{n=0}^{5} a_n T_n(x)$ corresponds to $1 - 2x + 3x^2 + 2x^3 + x^4 - x^5$.

9. Eigenvalues and Pseudospectra

Spectral methods are powerful tools for the computation of eigenvalues of differential and integral operators and their generalizations for strongly nonsymmetric problems, pseudospectra. Indeed, it was Orszag's 1971 computation of the critical Reynolds number $R = 5772.22$ for eigenvalue instability of plane Poiseuille fluid flow, a problem we shall discuss in Chapter 14, that did as much as anything to establish spectral methods as an important tool in scientific computing [Ors71]. Perhaps the reason why spectral methods are so important for eigenvalue computations is that these are applications where high accuracy tends to be crucial.

So far in this book we have seen two examples of eigenvalue calculations. Program 8 (p. 38) solved the harmonic oscillator problem

$$-u_{xx} + x^2 u = \lambda u, \qquad x \in \mathbb{R},$$

by a Fourier spectral method, taking advantage of the exponential decay of the eigenfunctions to replace the real line $\mathbb{R}$ by the periodic interval $[-L, L]$. Program 15 (p. 66) solved the even simpler problem

$$u_{xx} = \lambda u, \qquad -1 < x < 1, \quad u(\pm 1) = 0$$

by a Chebyshev spectral method that imposed the homogeneous Dirichlet conditions explicitly. In this chapter, we will develop such methods further with the aid of four additional examples. In each case we apply spectral ideas via matrices rather than the FFT, since it is so convenient to take advantage of the standard powerful algorithms for matrix eigenvalue and generalized eigenvalue problems embodied in the MATLAB commands eig and eigs.

Our four examples and the special features they illustrate can be summarized as follows:

Program 21: Mathieu equation, periodic domain;
Program 22: Airy equation, generalized eigenvalue problem;
Program 23: membrane oscillations, two-dimensional domain;
Program 24: complex harmonic oscillator, pseudospectra.

We begin with the Mathieu equation, an ODE that arises in problems of forced oscillations. (This example has been taken from Weideman and Reddy [WeRe00].) The equation can be written

$$-u_{xx} + 2q\cos(2x)u = \lambda u, \qquad (9.1)$$

where q is a real parameter, and we look for periodic solutions on $[-\pi, \pi]$. For $q = 0$, we have the linear pendulum equation of Program 15, with eigenvalues $n^2/4$ for $n = 1, 2, 3, \ldots$. The scientific interest arises in the behavior of these eigenvalues as q is increased.

To compute eigenvalues of the Mathieu equation by a spectral method, Program 21 discretizes (9.1) in a routine fashion. Translating the equation from the domain $[-\pi, \pi]$ to $[0, 2\pi]$ leaves the eigenvalues unaltered, so our discretization takes the form

$$L_N = -D_N^{(2)} + 2q\,\text{diag}(\cos(2x_1), \ldots, \cos(2x_N)),$$

where $D_N^{(2)}$ is the second-order Fourier differentiation matrix. The computation is straightforward, and with $N = 42$, we get about 13 digits of accuracy. Output 21 presents the plot generated by this program, showing the curves traced by the first 11 eigenvalues as q increases from 0 to 15. Producing this image took about half a second on my workstation. As shown in the figure, it is almost identical to Figure 20.1 on p. 724 of the classic *Handbook of Mathematical Functions* published by Abramowitz and Stegun in the 1960s [AbSt65]. (A few imperfections in the *Handbook* plot can be discerned, for example in the slope of the a_2 curve at $q = 0$.) We do not know how many seconds it took Gertrude Blanch, the author of the chapter on Mathieu functions in the *Handbook*, to produce that figure.

For our second example we turn to another classical problem of applied mathematics, the Airy equation [AbSt65, BeOr78]. Traditionally, the Airy equation is posed on the real line,

$$u_{xx} = xu, \qquad x \in \mathbb{R}. \qquad (9.2)$$

This is the canonical example of an ODE that changes type in different parts of the domain. For $x < 0$, the behavior is oscillatory, while for $x > 0$, we get growing and decaying exponential solutions. Being an ODE of second order, the Airy equation has a two-dimensional linear space of solutions, and the

9. Eigenvalues and Pseudospectra

Program 21

```
% p21.m - eigenvalues of Mathieu operator -u_xx + 2qcos(2x)u
%         (compare p8.m and p. 724 of Abramowitz & Stegun)
  N = 42; h = 2*pi/N; x = h*(1:N);
  D2 = toeplitz([-pi^2/(3*h^2)-1/6 ...
              -.5*(-1).^(1:N-1)./sin(h*(1:N-1)/2).^2]);
  qq = 0:.2:15; data = [];
  for q = qq;
    e = sort(eig(-D2 + 2*q*diag(cos(2*x))))';
    data = [data; e(1:11)];
  end
  clf, subplot(1,2,1)
  set(gca,'colororder',[0 0 1],'linestyleorder','-|--'), hold on
  plot(qq,data), xlabel q, ylabel \lambda
  axis([0 15 -24 32]), set(gca,'ytick',-24:4:32)
```

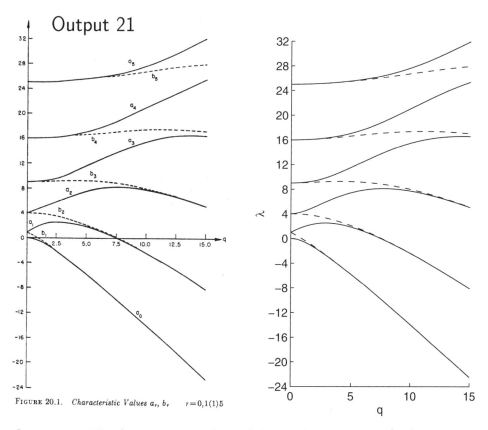

Output 21: *The first 11 eigenvalues of the Mathieu equation* (9.1). *Left, Figure 20.1 from Abramowitz and Stegun* (1965). *Right, output from Program* 21.

standard basis for this space is the pair of Airy functions Ai(x), which decays exponentially as $x \to \infty$, and Bi(x), which grows exponentially.

To solve the Airy equation numerically on the real line by spectral methods is not entirely straightforward, because there are an infinite number of oscillations that decay only algebraically in amplitude; there is an essential singularity at $-\infty$. Instead, in Program 22 we consider a slightly different eigenvalue problem posed on a finite interval,

$$u_{xx} = \lambda x u, \quad u(\pm 1) = 0, \quad -1 < x < 1. \tag{9.3}$$

This differs from our previous eigenvalue problems in a basic way: rather than being of the form (in linear algebra notation) $Au = \lambda u$, it is a *generalized eigenvalue problem* of the form $Au = \lambda Bu$. If B is nonsingular, then a generalized eigenvalue problem can be reduced mathematically to a standard one, $B^{-1}Au = \lambda u$. However, this is not necessarily a good idea in practice, and if B is singular, it is impossible even in principle. Instead, alternative numerical methods are generally used that deal with the generalized problem directly. The most standard is known as the QZ algorithm [GoVa96]. In MATLAB one calls upon the QZ algorithm by writing `eig(A,B)` instead of `eig(A)`. We shall not give details.

To discretize (9.3), we use a standard Chebyshev formulation for the second derivative and a diagonal matrix for the pointwise multiplication:

$$Au = \lambda Bu, \quad A = \tilde{D}_N^2, \quad B = \text{diag}(x_0, \ldots, x_N).$$

The computation is straightforward, and it is evident from Output 22 that we have spectral convergence, with ten or more digits of accuracy in the fifth eigenvector.

Figure 9.1 makes the connection back to Airy functions. In solving (9.3) we have imposed the condition $u(-1) = 0$ and looked for the fifth positive eigenvalue. This is equivalent to computing Ai(x) on the interval $[-L, L]$, where $-L = -7.944133\ldots$ is the location of the fifth zero of Ai(x). A rescaling back to $[-1, 1]$ then introduces a power $L^3 = 501.348\ldots$, and that is why our eigenvalue came out as L^3. Actually, what we have just said is not exactly true, for it assumed Ai($+L$) = 0, whereas in fact, Ai(L) $\simeq 5.5 \times 10^{-8}$. What we have computed is actually a spectrally accurate approximation of a linear combination of Ai(Lx) plus a very small multiple, of order 10^{-14}, of Bi(Lx). The coefficient 10^{-14} arises because Bi(L) $\simeq 10^6$.

Our third example is a Laplace eigenvalue problem in two space dimensions,

$$-\Delta u + f(x,y)u = \lambda u, \quad -1 < x, y < 1, \quad u = 0 \text{ on the boundary}. \tag{9.4}$$

For $f = 0$, we have a familiar problem easily solvable by separation of variables: the eigenfunctions have the form

$$\sin(k_x(x+1))\sin(k_y(y+1)),$$

9. Eigenvalues and Pseudospectra

Program 22

```
% p22.m - 5th eigenvector of Airy equation u_xx = lambda*x*u
  clf
  for N = 12:12:48
    [D,x] = cheb(N); D2 = D^2; D2 = D2(2:N,2:N);
    [V,Lam] = eig(D2,diag(x(2:N)));    % generalized ev problem
    Lam = diag(Lam); ii = find(Lam>0);
    V = V(:,ii); Lam = Lam(ii);
    [foo,ii] = sort(Lam); ii = ii(5); lambda = Lam(ii);
    v = [0;V(:,ii);0]; v = v/v(N/2+1)*airy(0);
    xx = -1:.01:1; vv = polyval(polyfit(x,v,N),xx);
    subplot(2,2,N/12), plot(xx,vv), grid on
    title(sprintf('N = %d      eig = %15.10f',N,lambda))
  end
```

Output 22

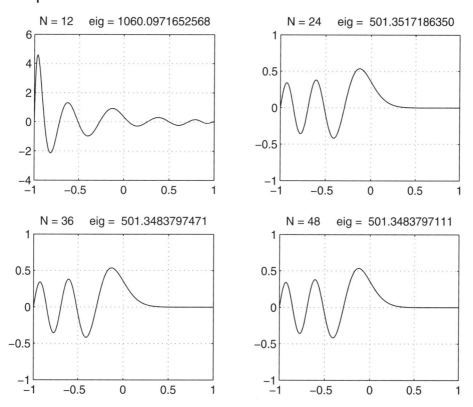

Output 22: *Convergence to the fifth eigenvector of the Airy problem* (9.3).

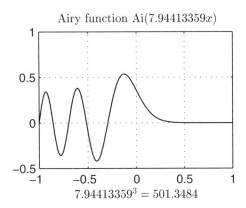

Fig. 9.1. *A rescaled solution of the Airy equation,* $\mathrm{Ai}(\lambda^{1/3}x)$. *This differs from the solution of Output 22 by about* 10^{-8}.

where k_x and k_y are integer multiples of $\pi/2$. This gives eigenvalues

$$\frac{\pi^2}{4}(i^2 + j^2), \qquad i,j = 1,2,3,\ldots.$$

Note that most of the eigenvalues are degenerate: whenever $i \neq j$, the eigenvalue has multiplicity 2. For $f \neq 0$, on the other hand, (9.4) will have no analytic solution in general and the eigenvalues will not be degenerate. Perturbations will split the double eigenvalues into pairs, a phenomenon familiar to physicists.

To solve (9.4) numerically by a spectral method, we can proceed just as in Program 16 (p. 70). We again set up the discrete Laplacian (7.5) of dimension $(N-1)^2 \times (N-1)^2$ as a Kronecker sum. To this we add a diagonal matrix consisting of the perturbation f evaluated at each of the $(N-1)^2$ points of the grid in the lexicographic ordering described on p. 68. The result is a large matrix whose eigenvalues can be found by standard techniques. In Program 23, this is done by MATLAB's command `eig`. For large enough problems, it would be important to use instead a Krylov subspace iterative method such as the Arnoldi or (if the matrix is symmetric) Lanczos iterations, which are implemented within MATLAB in the alternative code `eigs` (Exercise 9.4).

Output 23a shows results from Program 23 for the unperturbed case, computed by executing the code exactly as printed except with the line `L = L + diag(...)` commented out. Contour plots are given of the first four eigenmodes, with eigenvalues equal to $\pi^2/4$ times 2, 5, 5, and 8. As predicted, two of the eigenmodes are degenerate. As always in cases of degenerate eigenmodes, the choice of eigenvectors here is arbitrary. For essentially arbitrary reasons, the computation picks an eigenmode with a nodal line approximately along a diagonal; it then computes a second eigenmode linearly independent

9. Eigenvalues and Pseudospectra

Program 23

```
% p23.m - eigenvalues of perturbed Laplacian on [-1,1]x[-1,1]
%          (compare p16.m)

% Set up tensor product Laplacian and compute 4 eigenmodes:
  N = 16; [D,x] = cheb(N); y = x;
  [xx,yy] = meshgrid(x(2:N),y(2:N)); xx = xx(:); yy = yy(:);
  D2 = D^2; D2 = D2(2:N,2:N); I = eye(N-1);
  L = -kron(I,D2) - kron(D2,I);               % Laplacian
  L = L + diag(exp(20*(yy-xx-1)));            % + perturbation
  [V,D] = eig(L); D = diag(D);
  [D,ii] = sort(D); ii = ii(1:4); V = V(:,ii);

% Reshape them to 2D grid, interpolate to finer grid, and plot:
  [xx,yy] = meshgrid(x,y);
  fine = -1:.02:1; [xxx,yyy] = meshgrid(fine,fine);
  uu = zeros(N+1,N+1);
  [ay,ax] = meshgrid([.56 .04],[.1 .5]); clf
  for i = 1:4
    uu(2:N,2:N) = reshape(V(:,i),N-1,N-1);
    uu = uu/norm(uu(:),inf);
    uuu = interp2(xx,yy,uu,xxx,yyy,'cubic');
    subplot('position',[ax(i) ay(i) .38 .38])
    contour(fine,fine,uuu,-.9:.2:.9)
    colormap([0 0 0]), axis square
    title(['eig = ' num2str(D(i)/(pi^2/4),'%18.12f') '\pi^2/4'])
  end
```

of the first (though not orthogonal to it), with a nodal line approximately on the opposite diagonal. An equally valid pair of eigenmodes in this degenerate case would have had nodal lines along the x and y axes.

A remarkable feature of Output 23a is that although the grid is only of size 16×16, the eigenvalues are computed to 12-digit accuracy. This reflects the fact that one or two oscillations of a sine wave can be approximated to better than 12-digit precision by a polynomial of degree 16 (Exercise 9.1).

Output 23b presents the same plot with the perturbation in (9.4) included, with

$$f(x,y) = \exp(20(y - x - 1)).$$

This perturbation has a very special form. It is nearly zero outside the upper left triangular region, one-eighth of the total domain, defined by $y - x \geq 1$. Within that region, however, it is very large, achieving values as great as

Output 23a

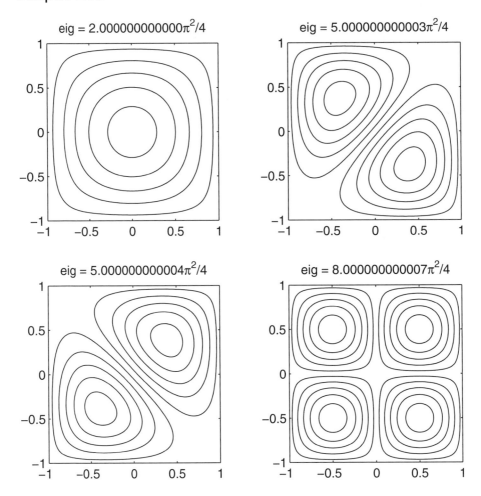

Output 23a: *First four eigenmodes of the Laplace problem (9.4) with $f(x,y) = 0$. These plots were produced by running Program 23 with the "+ perturbation" line commented out.*

4.8×10^8. Thus this perturbation is not small at all in amplitude, though it is limited in extent. It is analogous to the "barrier functions" utilized in the field of optimization of functions with constraints. The effect on the eigenmodes is clear. In Output 23b we see that all four eigenmodes avoid the upper left corner; the values there are very close to zero. It is approximately as if we had solved the eigenvalue problem on the unit square with a corner snipped off. All four eigenvalues have increased, as they must, and the second and third eigenvalues are no longer degenerate. What we find instead is that mode 3, which had low amplitude in the barrier region, has changed a little, whereas

9. Eigenvalues and Pseudospectra

Output 23b

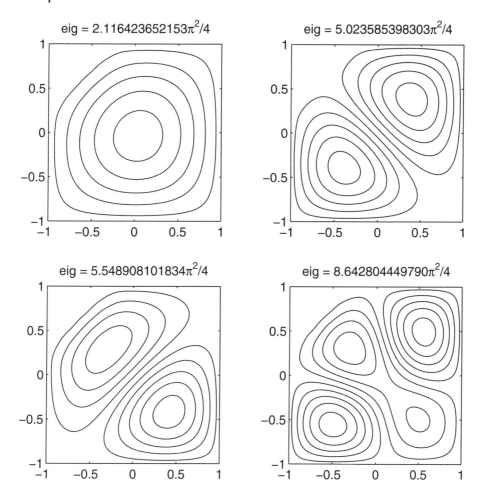

Output 23b: *First four eigenmodes of the perturbed Laplace problem* (9.4) *with* $f(x,y) = \exp(20(y-x-1))$. *These plots were produced by running Program 23 as written.*

mode 2, which had higher amplitude there, has changed quite a lot. These computed eigenvalues, by the way, are not spectrally accurate; the function f varies too fast to be well resolved on this grid. Experiments with various values of N suggest they are accurate to about three or four digits.

All of our examples of eigenvalue problems so far have involved self-adjoint operators, whose eigenvalues are real and whose eigenvectors can be taken to be orthogonal. Our spectral discretizations are not in fact symmetric matrices (they would be, if we used certain Galerkin rather than collocation methods), but they are reasonably close in the sense of having eigenvectors reasonably close to orthogonal so long as the corresponding eigenvalues are distinct. In general, a matrix with a complete set of orthogonal eigenvectors is said to be *normal*. Normal and nearly normal matrices are the ones whose eigenvalue problems are unproblematic, relatively easy both to solve and to interpret physically.

In a certain minority of applications, however, one encounters matrices or operators that are very far from normal in the sense that the eigenvectors, if a complete set exists, are very far from orthogonal—they form an ill-conditioned basis of the vector space under study. In highly nonnormal cases, it may be informative to compute pseudospectra* rather than spectra [Tre97, TTRD93, Wri00]. Suppose that a square matrix A is given and $\|\cdot\|$ is a physically relevant norm. For each $\epsilon > 0$, the ϵ-*pseudospectrum* of A is the subset of the complex plane

$$\Lambda_\epsilon(A) = \{z \in \mathbb{C}: \|(zI - A)^{-1}\| \geq \epsilon^{-1}\}. \tag{9.5}$$

(We use the convention $\|(zI - A)^{-1}\| = \infty$ if z is an eigenvalue of A.) Alternatively, $\Lambda_\epsilon(A)$ can be characterized by eigenvalues of perturbed matrices:

$$\Lambda_\epsilon(A) = \{z \in \mathbb{C}: z \text{ is an eigenvalue of } A + E \text{ for some } E \text{ with } \|E\| \leq \epsilon\}. \tag{9.6}$$

If $\|\cdot\|$ is the 2-norm, as is convenient and physically appropriate in most applications (sometimes after a diagonal similarity transformation to get the scaling right), then a further equivalence is

$$\Lambda_\epsilon(A) = \{z \in \mathbb{C}: \sigma_{\min}(zI - A) \leq \epsilon\}, \tag{9.7}$$

where $\sigma_{\min}$ denotes the minimum singular value.

Pseudospectra can be computed by spectral methods very effectively, and our final example of this chapter illustrates this. The example returns to the harmonic oscillator (4.6), except that a complex coefficient c is now put in front of the quadratic term. We define our linear operator L by

$$Lu = -u_{xx} + cx^2 u, \qquad x \in \mathbb{R}. \tag{9.8}$$

The eigenvalues and eigenvectors for this problem are readily determined analytically: they are $\sqrt{c}\,(2k+1)$ and $\exp(-c^{1/2}x^2/2)H_k(c^{1/4}x)$ for $k = 0, 1, 2, \ldots$, where H_k is the kth Hermite polynomial [Exn83]. However, as E. B. Davies first noted [Dav99], the eigenmodes are exponentially far from orthogonal. Output 24 shows pseudospectra for (9.8) with $c = 1 + 3i$ computed in a

---

*Pseudospectra (plural of pseudospectrum) are sets in the complex plane; pseudospectral methods are spectral methods based on collocation, i.e., pointwise evaluations rather than integrals. There is no connection—except that pseudospectral methods are very good at computing pseudospectra!

9. Eigenvalues and Pseudospectra

straightforward fashion based on (9.7). We discretize L spectrally, evaluate $\sigma_{\min}(zI - L)$ on a grid of points z_{ij}, then send the results to a contour plotter.

For the one and only time in this book, the plot printed as Output 24 is not exactly what would be produced by the corresponding program as listed. Program 24 evaluates $\sigma_{\min}(zI - L)$ on a relatively coarse 26×21 grid; after 546 complex singular value decompositions, a relatively crude approximation to Output 24 is produced. For the published figure, we made the grid four times finer in each direction by replacing 0:2:50 by 0:.5:50 and 0:2:40 by 0:.5:40. This slowed down the computation by a factor of 16. (As it happens, alternative algorithms can be used to speed up this calculation of pseudospectra and get approximately that factor of 16 back again; see [Tre99, Wri00].)

One can infer from Output 24 that although the eigenvalues of the complex harmonic oscillator are regularly spaced numbers along a ray in the complex plane, all but the first few of them would be of doubtful physical significance in a physical problem described by this operator. Indeed, the resolvent norm appears to grow exponentially as $|z| \to \infty$ along any ray with argument between 0 and $\arg c$, so that every value of z sufficiently far out in this infinite sector is an ϵ-pseudoeigenvalue for an exponentially small value of ϵ.

We shall see three further examples of eigenvalue calculations later in the book. We summarize the eigenvalue examples ahead by continuing the table displayed at the beginning of this chapter:

    Program 28:   circular membrane,   polar coordinates;
    Program 39:   square plate,   clamped boundary conditions;
    Program 40:   Orr–Sommerfeld operator,   complex arithmetic.

Summary of This Chapter. Spectral discretization can turn eigenvalue and pseudospectra problems for ODEs and PDEs into the corresponding problems for matrices. If the matrix dimension is large, it may be best to solve these by Krylov subspace methods such as the Lanczos or Arnoldi iterations.

Exercises

9.1. Modify Program 23 so that it produces a plot on a log scale of the error in the computed lowest eigenvalue represented in the first panel of Output 23a as a function of N. Now let $\tau > 0$ be fixed and let $E_N = \inf_p \|p(x) - \sin(\tau x)\|_\infty$, where $\|f\|_\infty = \sup_{x \in [-1,1]} |f(x)|$, denote the error in degree N minimax polynomial approximation to $\sin(\tau x)$ on $[-1, 1]$. It is known (see equation (6.77) of [Mei67]) that for even N, as $N \to \infty$, $E_N \sim 2^{-N} \tau^{N+1}/(N+1)!$. Explain which value of τ should be taken for this result to be used to provide an order of magnitude estimate of the results in the plot. How close is the estimate to the data? (Compare Exercise 5.3.)

Program 24

```
% p24.m - pseudospectra of Davies's complex harmonic oscillator
%         (For finer, slower plot, change 0:2 to 0:.5.)

% Eigenvalues:
  N = 70; [D,x] = cheb(N); x = x(2:N);
  L = 6; x = L*x; D = D/L;                    % rescale to [-L,L]
  A = -D^2; A = A(2:N,2:N) + (1+3i)*diag(x.^2);
  clf, plot(eig(A),'.','markersize',14)
  axis([0 50 0 40]), drawnow, hold on

% Pseudospectra:
  x = 0:2:50; y = 0:2:40; [xx,yy] = meshgrid(x,y); zz = xx+1i*yy;
  I = eye(N-1); sigmin = zeros(length(y),length(x));
  h = waitbar(0,'please wait...');
  for j = 1:length(x), waitbar(j/length(x))
    for i = 1:length(y), sigmin(i,j) = min(svd(zz(i,j)*I-A)); end
  end, close(h)
  contour(x,y,sigmin,10.^(-4:.5:-.5)), colormap([0 0 0])
```

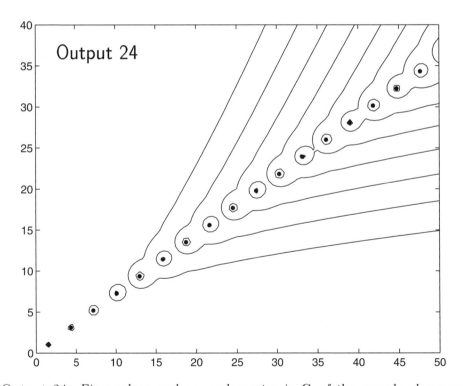

Output 24: *Eigenvalues and ϵ-pseudospectra in $\mathbb{C}$ of the complex harmonic oscillator* (9.8), $c = 1 + 3i$, $\epsilon = 10^{-0.5}, 10^{-1}, 10^{-1.5}, \ldots, 10^{-4}$.

9. Eigenvalues and Pseudospectra

9.2. $A \otimes (B \otimes C) = (A \otimes B) \otimes C$: True or false?

9.3. Modify Program 23 so that it finds the lowest eigenvalue of the Laplacian on the cube $[-1, 1]^3$ rather than the square $[-1, 1]^2$. For $N = 6$ and 8, how big is the matrix you are working with, how accurate are the results, and how long does the computation take? Estimate what the answers would be for $N = 12$.

9.4. In continuation of Exercise 9.3, you can solve the problem with $N = 12$ if you use MATLAB's iterative eigenvalue solver `eigs` rather than its "direct" solver `eig`. Modify your code further to use `eigs`, and be sure that `eigs` is given a sparse matrix to work with (putting `speye` instead of `eye` in your code will ensure this). With $N = 12$, how long does the computation take, and how accurate are the results?

9.5. Consider a circular membrane of radius 1 that vibrates according to the second-order wave equation $y_{tt} = r^{-1}(ry_r)_r + r^{-2}y_{\theta\theta}$, $y(1,t) = 0$, written in polar coordinates. Separating variables leads to consideration of solutions $y(r, \theta, t) = u(r)e^{im\theta}e^{i\omega t}$, with $u(r)$ satisfying $r^{-1}(ru_r)_r + (\omega^2 - r^{-2}m^2)u = 0$, $u_r(0) = 0$, $u(1) = 0$. This is a second-order, linear ODE boundary value problem with homogeneous boundary conditions, so one solution is $u(r) = 0$. Nonzero solutions will only occur for eigenvalues ω of the equation

$$r^{-1}(ru_r)_r - r^{-2}m^2 u = -\omega^2 u, \qquad u_r(0) = u(1) = 0. \tag{9.9}$$

This is a form of *Bessel's equation*, and the solutions are Bessel functions $J_m(\omega r)$, where ω has the property $J_m(\omega) = 0$. Write a MATLAB program based on a spectral method that, for given m, constructs a matrix whose smaller eigenvalues approximate the smaller eigenvalues of (9.9). (*Hint.* One method of implementing the Neumann boundary condition at $r = 0$ is described on p. 137.) List the approximations to the first six eigenvalues ω produced by your program for $m = 0, 1$ and $N = 5, 10, 15, 20$.

9.6. In continuation of Exercise 9.5, the first two eigenvalues for $m = 1$ differ nearly, but not quite, by a factor of 2. Suppose, with musical harmony in mind, we wish to design a membrane with radius-dependent physical properties such that these two eigenvalues have ratio exactly 2. Consider the modified boundary value eigenvalue problem

$$r^{-1}(p(r)ru_r)_r - r^{-2}m^2 u = -\omega^2 u, \qquad u_r(0) = u(1) = 0,$$

where $p(r) = 1 + \alpha \sin^2(\pi r)$ for some real number α. Produce a plot that shows the first eigenvalue and $\frac{1}{2}$ times the second eigenvalue as functions of α. For what value of α do the two curves intersect? By solving an appropriate nonlinear equation, determine this critical value of α to at least six digits. Can you explain why a correction of the form $\alpha p(r)$ modifies the ratio of the eigenvalues in the direction required?

9.7. Exercise 6.8 (p. 59) considered powers of the Chebyshev differentiation matrix D_N. For $N = 20$, produce a plot of the eigenvalues and ϵ-pseudospectra of D_N for $\epsilon = 10^{-2}, 10^{-3}, \ldots, 10^{-16}$. Comment on how this plot relates to the results of that exercise.

9.8. Download the MATLAB programs from [Wri00] for computing pseudospectra and use them to generate a figure similar to Output 24. How does the computation time compare to that of Program 24?

10. Time-Stepping and Stability Regions

When time-dependent PDEs are solved numerically by spectral methods, the pattern is usually the same: spectral differentiation in space, finite differences in time. For example, one might carry out the time-stepping by an Euler, leap frog, Adams, or Runge–Kutta formula [But87, HaWa96, Lam91]. In principle, one sacrifices spectral accuracy in doing so, but in practice, small time steps with formulas of order 2 or higher often leave the global accuracy quite satisfactory. Small time steps are much more affordable than small space steps, for they affect the computation time, but not the storage, and then only linearly. By contrast, halving the space step typically multiplies the storage by 2^d in d space dimensions, and it may multiply the computation time for each time step by anywhere from 2^d to 2^{3d}, depending on the linear algebra involved.

So far in this book we have solved three time-dependent PDEs, in each case by a leap frog discretization in t. The equations and the time steps we used were as follows:

p6: $u_t + c(x)u_x = 0$ on $[-\pi, \pi]$,   Fourier, $\Delta t = 1.57N^{-1}$;

p19: $u_{tt} = u_{xx}$ on $[-1, 1]$,   Chebyshev, $\Delta t = 8N^{-2}$;

p20: $u_{tt} = u_{xx} + u_{yy}$ on $[-1, 1]^2$,   2D Chebyshev, $\Delta t = 6N^{-2}$.

Now it is time to explain where these choices of Δt came from.

Figure 10.1 shows the output from Program 6 (p. 26) when the time step is increased to $\Delta t = 1.9N^{-1}$, and Figure 10.2 shows the output from Program 20 (p. 83) with $\Delta t = 6.6N^{-2}$. Catastrophes! Both computations are *numerically unstable* in the sense that small errors are amplified unboundedly—in fact,

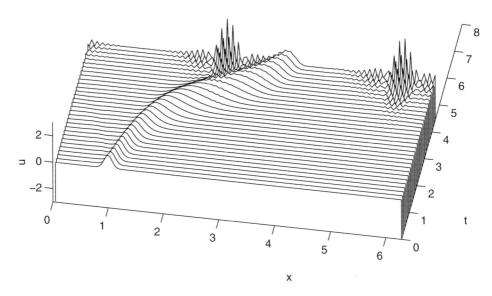

Fig. 10.1. *Repetition of Output 6 with $\Delta t = 1.9N^{-1}$. The time step is too large for stability, and sawtooth oscillations appear near $x = 1 + \pi/2$ and $1 + 3\pi/2$ that will grow exponentially and swamp the solution.*

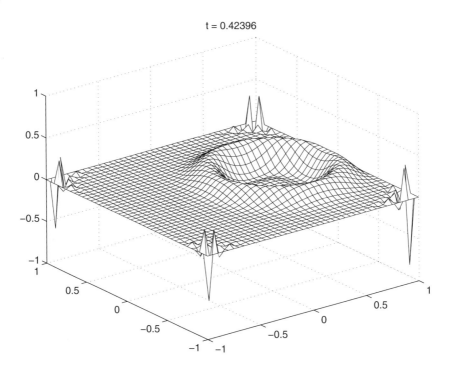

Fig. 10.2. *Repetition of Output 20 with $\Delta t = 6.6N^{-2}$. Again we have exponentially growing instability, with the largest errors at the corners.*

10. Time-Stepping and Stability Regions

exponentially. (With finite difference and finite element methods, it is almost always discretization errors that excite instability. With spectral methods the discretization errors are sometimes so small that rounding errors are important too.) In both cases, we have terminated the computation quickly after the instability sets in to make an attractive plot. Larger time steps or longer integrations easily lead to growth by many orders of magnitude and floating point overflow.

For Program 19 (p. 82) a similar instability appears with $\Delta t = 9.2N^{-2}$. After a little trial and error, we find that the stability restrictions are approximately as follows:

Program	Empirical stability restriction
p6	$\Delta t < 1.9N^{-1}$
p19	$\Delta t < 9.2N^{-2}$
p20	$\Delta t < 6.6N^{-2}$

The aim of this chapter is to show where such stability restrictions come from and to illustrate further that so long as they are satisfied—or circumvented by suitable implicit discretizations—spectral methods may be very powerful tools for time-dependent problems.

Many practical calculations can be handled by an analysis based on the notion of the *method of lines*. When a time-dependent PDE is discretized in space, whether by a spectral method or otherwise, the result is a coupled system of ODEs in time. The lines $x =$ constant are the "lines" alluded to in the name:

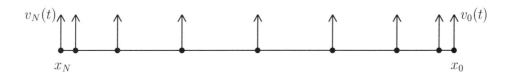

The method of lines refers to the idea of solving this coupled system of ODEs by a finite difference formula in t (Adams, Runge–Kutta, etc.). The rule of thumb for stability is as follows:

Rule of Thumb.

The method of lines is stable if the eigenvalues of the (linearized) spatial discretization operator, scaled by Δt, lie in the stability region of the time-discretization operator.

We hope that the reader is familiar with the notion of the *stability region* of an ODE formula. Briefly, it is the subset of the complex plane consisting of those $\lambda \in \mathbb{C}$ for which the numerical approximation produces bounded solutions when applied to the scalar linear model problem $u_t = \lambda u$ with time step Δt—multiplied by Δt, so as to make the scaling independent of Δt [But87, HaWa96, Lam91]. (For problems of second order in t, the model problem becomes $u_{tt}(t) = \lambda u(t)$ and one multiplies by $(\Delta t)^2$.)

The Rule of Thumb is not always reliable, and in particular, it may fail for problems involving discretization matrices that are far from normal, i.e., with eigenvectors far from orthogonal [TrTr87]. For such problems, the right condition is that the pseudospectra must lie within the stability region too: more precisely, the ϵ-pseudospectrum must lie within a distance $O(\epsilon) + O(\Delta t)$ of the stability region as $\epsilon \to 0$ and $\Delta t \to 0$ [KrWu93, ReTr92]. When in doubt about whether a discretization matrix is far from normal, it is a good idea to take a look at its pseudospectra, either by computing eigenvalues of a few randomly perturbed matrices or with the aid of a modification of Program 24 or the faster codes of [Tre99] and [Wri00]. For many problems, fortunately, the Rule of Thumb makes accurate predictions.

Program 25 plots various stability regions for standard Adams–Bashforth (explicit), Adams–Moulton (implicit), backward differentiation (implicit), and Runge–Kutta (explicit) formulas. Though we list the code as always, we will not discuss it at all but refer the reader to the textbooks cited above for explanations of how curves like these can be generated.

To analyze the time-stepping in Programs 6, 19, and 20, we need the stability region for the leap frog formula, which is not covered by Program 25. For $u_t = \lambda u$, the leap frog formula is

$$\frac{v^{(n+1)} - v^{(n-1)}}{2\Delta t} = \lambda v^{(n)}. \tag{10.1}$$

The characteristic equation for this recurrence relation is $g - g^{-1} = 2\lambda\Delta t$, which we obtain by inserting in (10.1) the ansatz $v^{(n)} = g^n$, and the condition for stability is that both roots of this equation must lie in the closed unit disk, with only simple roots permitted on the unit circle. Now it is clear that if g is one root, then $-g^{-1}$ is the other. If $|g| < 1$, then $|-g^{-1}| > 1$, giving instability. Thus stability requires $|g| = 1$ and $g \neq -g^{-1}$, hence $g \neq \pm i$. That is, stable values of g range over the unit circle except for $\pm i$, and the corresponding values of $g - g^{-1}$ fill the open complex interval $(-2i, 2i)$. We conclude that the leap frog formula applied to $u_t = \lambda u$ is stable provided $2\lambda\Delta t$ belongs to $(-2i, 2i)$; i.e., the stability region in the $\lambda\Delta t$-plane is $(-i, i)$ (Figure 10.3).

Let us apply this conclusion to the "frozen coefficient" analogue of the PDE of Program 6, $u_t + u_x = 0$. By working in the Fourier domain we see that the eigenvalues of the Fourier spectral differentiation matrix D_N are the numbers ik for $k = -N/2 + 1, \ldots, N/2 - 1$, with zero having multiplicity 2. Thus the

10. Time-Stepping and Stability Regions

Program 25

```
% p25.m - stability regions for ODE formulas

% Adams-Bashforth:
  clf, subplot('position',[.1 .56 .38 .38])
  plot([-8 8],[0 0]), hold on, plot([0 0],[-8 8])
  z = exp(1i*pi*(0:200)/100); r = z-1;
  s = 1; plot(r./s)                                    % order 1
  s = (3-1./z)/2; plot(r./s)                           % order 2
  s = (23-16./z+5./z.^2)/12; plot(r./s)                % order 3
  axis([-2.5 .5 -1.5 1.5]), axis square, grid on
  title Adams-Bashforth

% Adams-Moulton:
  subplot('position',[.5 .56 .38 .38])
  plot([-8 8],[0 0]), hold on, plot([0 0],[-8 8])
  s = (5*z+8-1./z)/12; plot(r./s)                      % order 3
  s = (9*z+19-5./z+1./z.^2)/24; plot(r./s)             % order 4
  s = (251*z+646-264./z+106./z.^2-19./z.^3)/720; plot(r./s)   % 5
  d = 1-1./z;
  s = 1-d/2-d.^2/12-d.^3/24-19*d.^4/720-3*d.^5/160; plot(d./s) % 6
  axis([-7 1 -4 4]), axis square, grid on, title Adams-Moulton

% Backward differentiation:
  subplot('position',[.1 .04 .38 .38])
  plot([-40 40],[0 0]), hold on, plot([0 0],[-40 40])
  r = 0; for i = 1:6, r = r+(d.^i)/i; plot(r), end    % orders 1-6
  axis([-15 35 -25 25]), axis square, grid on
  title('backward differentiation')

% Runge-Kutta:
  subplot('position',[.5 .04 .38 .38])
  plot([-8 8],[0 0]), hold on, plot([0 0],[-8 8])
  w = 0; W = w; for i = 2:length(z)                    % order 1
    w = w-(1+w-z(i)); W = [W; w]; end, plot(W)
  w = 0; W = w; for i = 2:length(z)                    % order 2
    w = w-(1+w+.5*w^2-z(i)^2)/(1+w); W = [W; w];
    end, plot(W)
  w = 0; W = w; for i = 2:length(z)                    % order 3
    w = w-(1+w+.5*w^2+w^3/6-z(i)^3)/(1+w+w^2/2); W = [W; w];
    end, plot(W)
  w = 0; W = w; for i = 2:length(z)                    % order 4
    w = w-(1+w+.5*w^2+w^3/6+w.^4/24-z(i)^4)/(1+w+w^2/2+w.^3/6);
    W = [W; w]; end, plot(W)
  axis([-5 2 -3.5 3.5]), axis square, grid on, title Runge-Kutta
```

Output 25

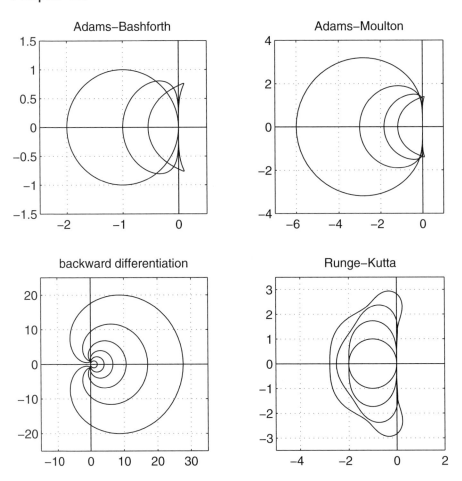

Output 25: *Stability regions for four families of ODE finite difference formulas. For backward differentiation, the stability regions are the exteriors of the curves; in the other cases they are the interiors.*

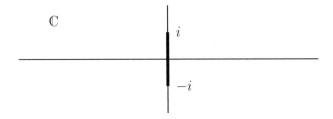

Fig. 10.3. *Stability region of the leap frog formula* (10.1) *for a first derivative.*

10. Time-Stepping and Stability Regions

stability condition for Fourier spectral discretization in space coupled with the leap frog formula in time for $u_t = u_x$ on $[-\pi, \pi]$ is

$$\Delta t \left(\tfrac{1}{2}N - 1\right) < 1,$$

that is, approximately $\Delta t \leq 2N^{-1}$. If we were to increase Δt gradually across this threshold, the first modes to go unstable would be of the form $e^{\pm i(N/2-1)x_j}$, that is, approximately sawtooths.

Now in Program 6, we have the equation $u_t + c(x)u_x = 0$, where c is a variable coefficient that takes a maximum of $6/5$ at $x = 1 + \pi/2$ and $1 + 3\pi/2$. For large N, the largest eigenvalues will accordingly be about $6/5$ times larger than in the analysis just carried out for $u_t = u_x$. This gives the approximate stability condition

$$\Delta t \leq \tfrac{5}{3}N^{-1}.$$

This condition is slightly stricter than the observed $1.9N^{-1}$; the agreement would be better for larger N (Exercise 10.1). Note that it is precisely at the parts of the domain where c is close to its maximum that the instability first appears in Output 6, and that it has the predicted form of an approximate sawtooth.

For Programs 19 and 20, we have the leap frog approximation of a second derivative. This means we have a new stability region to figure out. Applying the leap frog formula to the model problem $u_{tt} = \lambda u$ gives

$$\frac{v^{(n+1)} - 2v^{(n)} + v^{(n-1)}}{(\Delta t)^2} = \lambda v^{(n)}. \tag{10.2}$$

The characteristic equation of this recurrence relation is $g + g^{-1} = \lambda(\Delta t)^2 + 2$, and if g is one root, the other is g^{-1}. By a similar calculation as before, we deduce that the stability region in the $\lambda(\Delta t)^2$-plane is the real negative open interval $(-4, 0)$ (Figure 10.4).

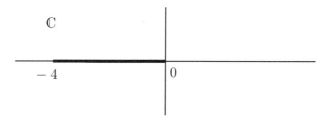

Fig. 10.4. *Stability region of the leap frog formula (10.2) for a second derivative.*

According to the Rule of Thumb, for a given spectral discretization, we must pick Δt small enough that this stability region encloses the eigenvalues of the spectral discretization operator, scaled by Δt. For Program 19, the spatial discretization operator is $\widetilde{D}_N^2$. The eigenvalues of $\widetilde{D}_N^2$ (we shall give details in a moment) are negative real numbers, the largest of which in magnitude is approximately $-0.048N^4$. For Program 19, accordingly, our stability restriction is approximately $-0.048N^4(\Delta t)^2 \leq -4$, i.e.,

$$\Delta t \leq 9.2N^{-2},$$

and when this condition is violated, trouble should arise first at the boundaries, where the offending eigenmodes are concentrated. These predictions match observations.

Program 20, in two dimensions, is easily seen to have largest eigenvalues approximately twice as large as in Program 19. This means that the stability condition is twice as strict on $(\Delta t)^2$, hence $\sqrt{2}$ times as strict on Δt,

$$\Delta t \leq 6.5N^{-2}.$$

Again, this estimate matches observations, and our analysis explains why the oscillations in Output 20 appeared in the corners of the domain.

We have just asserted that the eigenvalues of $\widetilde{D}_N^2$ are negative and real, with the largest being approximately $-0.048N^4$. There is a great deal to be said about this matrix, and in fact, we have already considered it in Program 15 (p. 66). First of all, it is noteworthy that although $\widetilde{D}_N^2$ approximates the Hermitian operator d^2/dx^2 with appropriate boundary conditions on $[-1, 1]$, it is nonsymmetric. Nonetheless, the eigenvalues have been proved to be real [GoLu83b], and many of them are spectrally accurate approximations to the eigenvalues $-k^2\pi^2/4$ of d^2/dx^2. As $N \to \infty$, the fraction of eigenvalues that behave in this way converges to $2/\pi$ [WeTr88]. The explanation for this number is that in the center of the grid, where the spacing is coarsest, the highest wavenumber of a sine function for which there are at least two grid points per wavelength is $2N/\pi$. Now what about the remaining eigenvalues of $\widetilde{D}_N^2$, with proportion $1 - 2/\pi$ asymptotically as $N \to \infty$? These turn out to be very large, of order N^4, and physically meaningless. They are called *outliers*, and the largest in magnitude is about $-0.048N^4$ [Van90].

Program 26 calculates the eigenvalues of $\widetilde{D}_N^2$ and plots one of the physically meaningful eigenvectors and one of the physically meaningless ones. We have already had a taste of this behavior with Program 15. These outliers correspond to nonphysical eigenmodes that are not global sines and cosines, but strongly localized near $x = \pm 1$.

We complete this chapter by solving another time-dependent PDE, this time a famous nonlinear one, by a spectral method involving a Runge–Kutta discretization in time. The *KdV equation* takes the form

$$u_t + uu_x + u_{xxx} = 0, \tag{10.3}$$

10. Time-Stepping and Stability Regions

Program 26

```
% p26.m - eigenvalues of 2nd-order Chebyshev diff. matrix

  N = 60; [D,x] = cheb(N); D2 = D^2; D2 = D2(2:N,2:N);
  [V,Lam] = eig(D2);
  [foo,ii] = sort(-diag(Lam)); e = diag(Lam(ii,ii)); V = V(:,ii);

% Plot eigenvalues:
  clf, subplot('position',[.1 .62 .8 .3])
  loglog(-e,'.','markersize',10), ylabel eigenvalue
  title(['N = ' int2str(N) ...
   ',      max |\lambda| = ' num2str(max(-e)/N^4) 'N^4'])
  hold on, semilogy(2*N/pi*[1 1],[1 1e6],'--r')
  text(2.1*N/pi,24,'2\pi / N','fontsize',12)

% Plot eigenmodes N/4 (physical) and N (nonphysical):
  vN4 = [0; V(:,N/4-1); 0];
  xx = -1:.01:1; vv = polyval(polyfit(x,vN4,N),xx);
  subplot('position',[.1 .36 .8 .15]), plot(xx,vv), hold on
  plot(x,vN4,'.','markersize',9), title('eigenmode N/4')
  vN = V(:,N-1);
  subplot('position',[.1 .1 .8 .15])
  semilogy(x(2:N),abs(vN)), axis([-1 1 5e-6 1]), hold on
  plot(x(2:N),abs(vN),'.','markersize',9)
  title('absolute value of eigenmode N    (log scale)')
```

a blend of a nonlinear hyperbolic term uu_x and a linear dispersive term u_{xxx}. Among the solutions admitted by (10.3) are *solitary waves*, traveling waves of the form

$$u(x,t) = 3a^2 \operatorname{sech}^2\bigl(a(x-x_0)/2 - a^3 t\bigr) \qquad (10.4)$$

for any real a and x_0. (Here sech denotes the inverse of the hyperbolic cosine, $\operatorname{sech}(x) = 2/(e^x + e^{-x})$.) Note that this wave has amplitude $3a^2$ and speed $2a^2$, so the speed is proportional to the amplitude. This is in contrast to linear wave equations, where the speed is independent of the amplitude. Also, note that the value of u decays rapidly in space away from $x = x_0 + 2a^2 t$, so the waves are localized in space.

What is most remarkable about (10.3) is that solutions exist that consist almost exactly of finite superpositions of waves (10.4) of arbitrary speeds that interact cleanly, passing through one another with the only lasting effect of the interaction being a phase shift of the individual waves. These interacting solitary waves are called *solitons*, and their behavior has been a celebrated

Output 26

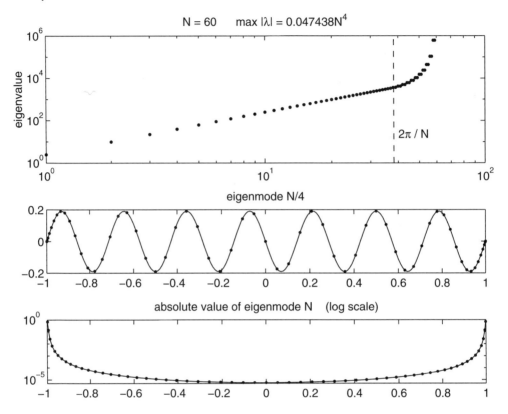

Output 26: *The top plot shows the sorted eigenvalues of $\widetilde{D}_N^2$. A fraction approximately $2/\pi$ of them correspond to good approximations of the sinusoidal eigenmodes of $u_{xx} = \lambda u$, $u(\pm 1) = 0$. Mode $N/4$ is one such, whereas mode N is spurious and localized near the boundaries—note the log scale.*

topic of applied mathematics since the 1970s [DrJo89, Whi74]. One of the most striking early applications of spectral methods was the computation of interacting solitons by Fornberg and Whitham described in a classic article in 1978 [FoWh78].

Program 27 models the KdV equation by a Fourier spectral method on $[-\pi, \pi]$, which is appropriate since we are not interested in the effect of boundary conditions and the solutions at issue decay exponentially. This is the first nonlinear time-dependent equation we have considered, but the nonlinearity causes little trouble for an explicit time-stepping method. The time-discretization scheme in this program is the fourth-order Runge–Kutta formula, which is described in numerous books.

If Program 27 were written in the obvious manner, it would compute solutions successfully but would need a very small time step for stability. Instead, the code is constructed in a modified form, based on the *method of integrating factors*, which allows time steps five or ten times larger. The initial condition is the superposition of two solitons. The solitons pass through each other as expected with only a change in phase. This computation involves 983 time steps and takes about ten seconds on my workstation.

The method of integrating factors is based on the idea that the problem can be transformed so that the linear part of the PDE is solved exactly [ChKe85, MiTa99]. Since the linear term in (10.3) is the one involving high frequencies, the "stiff" term that constrains the stability, this leads to the possibility of larger time steps. One way to proceed is to write (10.3) as

$$u_t + (\tfrac{1}{2}u^2)_x + u_{xxx} = 0, \qquad (10.5)$$

with Fourier transform

$$\hat{u}_t + \tfrac{i}{2}k\widehat{u^2} - ik^3\hat{u} = 0.$$

Now we multiply by $e^{-ik^3 t}$—this is the integrating factor—to get

$$e^{-ik^3 t}\hat{u}_t + \tfrac{i}{2}e^{-ik^3 t}k\widehat{u^2} - ie^{-ik^3 t}k^3\hat{u} = 0.$$

If we define $\hat{U} = e^{-ik^3 t}\hat{u}$, with $\hat{U}_t = -ik^3\hat{U} + e^{-ik^3 t}\hat{u}_t$, this becomes

$$\hat{U}_t + ik^3\hat{U} + \tfrac{i}{2}e^{-ik^3 t}k\widehat{u^2} - ik^3\hat{U} = 0,$$

i.e.,

$$\hat{U}_t + \tfrac{i}{2}e^{-ik^3 t}k\widehat{u^2} = 0.$$

The linear term is gone, and the problem is no longer stiff. (It now has a rapidly varying coefficient, however, so the improvement, while worthwhile, is not as great as one might have imagined.) Working in Fourier space, we can discretize the problem in the form

$$\hat{U}_t + \tfrac{i}{2}e^{-ik^3 t}k\mathcal{F}((\mathcal{F}^{-1}(e^{ik^3 t}\hat{U}))^2) = 0,$$

where $\mathcal{F}$ is the Fourier transform operator as in Exercise 2.1, and this is what is done by Program 27.

Summary of This Chapter. As a rule of thumb, stability of spectral methods for time-dependent PDEs requires that the eigenvalues of the spatial discretization operator, scaled by Δt, lie in the stability region of the time-stepping formula. Because of large eigenvalues, especially in the Chebyshev case, time

Program 27

```
% p27.m - Solve KdV eq. u_t + uu_x + u_xxx = 0 on [-pi,pi] by
%         FFT with integrating factor v = exp(-ik^3t)*u-hat.

% Set up grid and two-soliton initial data:
  N = 256; dt = .4/N^2; x = (2*pi/N)*(-N/2:N/2-1)';
  A = 25; B = 16; clf, drawnow
  u = 3*A^2*sech(.5*(A*(x+2))).^2 + 3*B^2*sech(.5*(B*(x+1))).^2;
  v = fft(u); k = [0:N/2-1 0 -N/2+1:-1]'; ik3 = 1i*k.^3;

% Solve PDE and plot results:
  tmax = 0.006; nplt = floor((tmax/25)/dt); nmax = round(tmax/dt);
  udata = u; tdata = 0; h = waitbar(0,'please wait...');
  for n = 1:nmax
    t = n*dt; g = -.5i*dt*k;
    E = exp(dt*ik3/2); E2 = E.^2;
    a = g.*fft(real( ifft(     v     ) ).^2);
    b = g.*fft(real( ifft(E.*(v+a/2)) ).^2);    % 4th-order
    c = g.*fft(real( ifft(E.*v + b/2) ).^2);    % Runge-Kutta
    d = g.*fft(real( ifft(E2.*v+E.*c) ).^2);
    v = E2.*v + (E2.*a + 2*E.*(b+c) + d)/6;
    if mod(n,nplt) == 0
      u = real(ifft(v)); waitbar(n/nmax)
      udata = [udata u]; tdata = [tdata t];
    end
  end
  waterfall(x,tdata,udata'), colormap([0 0 0]), view(-20,25)
  xlabel x, ylabel t, axis([-pi pi 0 tmax 0 2000]), grid off
  set(gca,'ztick',[0 2000]), close(h), pbaspect([1 1 .13])
```

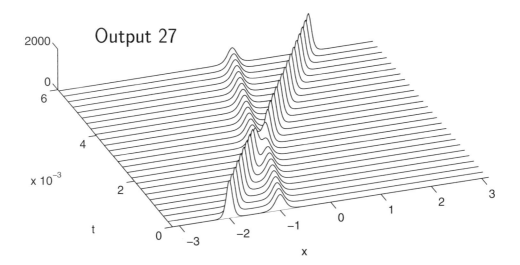

Output 27

10. Time-Stepping and Stability Regions

step limits for explicit methods may be very severe, making it advantageous to use implicit or semi-implicit methods.

Exercises

10.1. In our stability analysis of Program 6, we "froze the coefficients" and assumed that the largest eigenvalue of the discretization of $u_t + c(x)u_x = 0$ would be about 6/5 times that of the discretization of $u_t + u_x$. Perform a numerical study to investigate how true this is. (You may wish to work with a matrix formulation of the algorithm, as in Exercise 3.7.) Produce a plot of the ratio of the actual and estimated eigenvalues for $N = 20, 40, 60, \ldots, 200$. For $N = 128$, how does the true eigenvalue compare with the frozen-coefficient prediction? What stability restriction does the true eigenvalue suggest? Does this match the empirically observed stability restriction? Does the corresponding eigenvector look like the unstable mode visible in Figure 10.1?

10.2. Rerun Program 27 with the time step increased from $0.4N^{-2}$ to $0.45N^{-2}$. Comment on the resulting plot. Can you explain this effect with reference to stability regions?

10.3. Consider the first-order linear initial boundary value problem

$$u_t = u_x, \quad x \in [-1, 1], \quad 0 < t < 1, \quad u(1, t) = 0,$$

with initial data $u(x, 0) = \exp(-60(x - 1/2)^2)$. Write a program to solve this problem by a matrix-based Chebyshev spectral discretization in x coupled with the third-order Adams–Bashforth formula in t, for which the formula is $v^{(n+3)} = v^{(n+2)} + \frac{1}{12}\Delta t(23f^{(n+2)} - 16f^{(n+1)} + 5f^{(n)})$. Initial values can be supplied from the exact solution. Take $N = 50$ and $\Delta t = \nu N^{-2}$, where ν is a parameter. For each of the two choices $\nu = 7$ and $\nu = 8$, produce one plot of the computed solution at $t = 1$ and another that superimposes the stability region in the $\lambda \Delta t$-plane, the eigenvalues of the spatial discretization matrix, and its ϵ-pseudospectra for $\epsilon = 10^{-2}, 10^{-3}, \ldots, 10^{-6}$. Comment on the results.

10.4. Consider the nonlinear initial boundary value problem

$$u_t = u_{xx} + e^u, \quad x \in [-1, 1], \quad t > 0, \quad u(\pm 1, t) = u(x, 0) = 0$$

for the unknown function $u(x, t)$. To at least eight digits of accuracy, what is $u(0, 3.5)$, and what is the time t_5 such that $u(0, t_5) = 5$?

10.5. Chebyshev grids have an $O(N^{-2})$ spacing near the boundaries. Therefore, it is sometimes said, it is obvious that an explicit Chebyshev spectral method for a hyperbolic PDE such as $u_t = u_x$ must require time steps of size $O(N^{-2})$, "because of the CFL (Courant–Friedrichs–Lewy) condition" [RiMo67]. Explain why this argument is invalid.

10.6. The KdV equation (10.3) is closely related to the *Burgers equation*, $u_t + (u^2)_x = \epsilon u_{xx}$, where $\epsilon > 0$ is a constant [Whi74]. Modify Program 27 to solve this equation for $\epsilon = 0.25$ by a Fourier spectral method on $[-\pi, \pi]$ with an integrating factor. Take $u(x,0)$ equal to $\sin^2(x)$ in $[-\pi, 0]$ and to zero in $[0, \pi]$, and produce plots at times $0, \frac{1}{2}, 1, \ldots, 3$, with a sufficiently small time step, for $N = 64$, 128, and 256. For $N = 256$, how small a value of ϵ can you take without obtaining unphysical oscillations?

10.7. Another related PDE is the *Kuramoto–Sivashinsky equation*, $u_t + (u^2)_x = -u_{xx} - u_{xxxx}$, whose solutions evolve chaotically. This equation is much more difficult to solve numerically. Write a program to solve it with periodic boundary conditions on the domain $[-20, 20]$ for initial data $u(x,0) = \exp(-x^2)$. Can you get results for $0 \leq t \leq 50$ that you trust?

10.8. Of course, the KdV equation is also applicable to initial data that do not consist of a simple superposition of solitons. Explore some of behaviors of this equation by modifying Program 27 to start with the initial function $u(x,0) = 1875\exp(-20x^2)$, as well as another function of your choosing.

11. Polar Coordinates

Spectral computations are frequently carried out in multidimensional domains in which one has different kinds of boundary conditions in the different dimensions. One of the most common examples is the use of *polar coordinates* in the unit disk,

$$x = r\cos\theta, \quad y = r\sin\theta.$$

Including a third variable z or ϕ would bring us to *cylindrical* or *spherical coordinates*.

The most common way to discretize the disk spectrally is to take a periodic Fourier grid in θ and a nonperiodic Chebyshev grid in r:

$$\theta \in [0, 2\pi], \quad r \in [0, 1].$$

Specifically, the grid in the r-direction is transformed from the usual Chebyshev grid for $x \in [-1, 1]$ by $r = (x+1)/2$. The result is a polar grid that is highly clustered near both the boundary and the origin, as illustrated in Figure 11.1. Grids like this are convenient and commonly used, but they have some drawbacks. One difficulty is that while it is sometimes advantageous to have points clustered near the boundary, it may be wasteful and is certainly inelegant to devote extra grid points to the very small region near the origin, if the solution is smooth there. Another is that for time-dependent problems, these small cells near the origin may force one to use excessively small time steps for numerical stability. Accordingly, various authors have found alternative ways to treat the region near $r = 0$. We shall describe one method of this kind in essentially the formulation proposed by Fornberg [For95, For96,

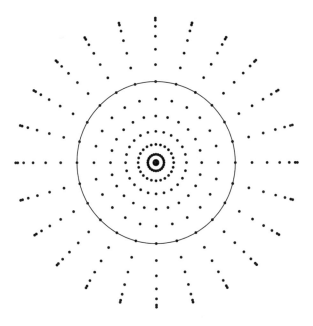

Fig. 11.1. *A spectral grid based on a Chebyshev discretization of $r \in [0,1]$. Half the grid points lie inside the circle, which encloses 31% of the total area.*

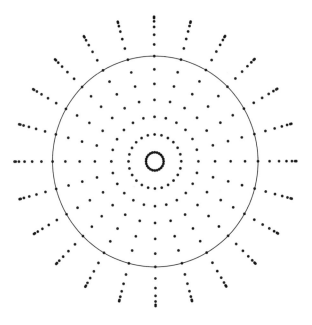

Fig. 11.2. *A spectral grid based on a Chebyshev discretization of $r \in [-1,1]$. Now the circle encloses 53% of the area.*

11. Polar Coordinates

FoMe97]. Closely related methods for polar and/or spherical coordinates have been used by others over the years; for a table summarizing 20 contributions in this area, see [Boy00].

The idea is to take $r \in [-1, 1]$ instead of $r \in [0, 1]$. To begin with, suppose θ continues to range over $[0, 2\pi]$. Then we have the coordinate system

$$\theta \in [0, 2\pi], \quad r \in [-1, 1], \tag{11.1}$$

illustrated in Figure 11.2. What is unusual about this representation is that each point (x, y) in the disk corresponds to two distinct points (r, θ) in coordinate space: the map from (r, θ) to (x, y) is 2-to-1. (At the special point $x = y = 0$, it is ∞-to-1, but we can avoid this complication by taking the grid parameter N in the r direction to be odd.) To put it another way, if a function $u(r, \theta)$ is to correspond to a single-valued function of x and y, then it must satisfy a symmetry condition in (r, θ)-space:

$$u(r, \theta) = u(-r, (\theta + \pi)(\bmod 2\pi)). \tag{11.2}$$

Once the condition (11.2) has been identified, it is not hard to implement it in a spectral method. To explain how this can be done, let us begin with a simplified variant of the problem. Suppose we want to compute a matrix-vector product Ax, where A is a $2N \times 2N$ matrix and x is a $2N$-vector. If we break A into four $N \times N$ blocks and x into two N-vectors, we can write the product in the form

$$Ax = \begin{array}{|c|c|} \hline A_1 & A_2 \\ \hline A_3 & A_4 \\ \hline \end{array} \begin{array}{|c|} \hline x_1 \\ \hline x_2 \\ \hline \end{array}. \tag{11.3}$$

Now suppose that we have the additional condition $x_1 = x_2$, and similarly, we know that the first N entries of Ax will always be equal to the last N entries. Then we have

$$(Ax)_{1:N} = (A_1 + A_2)x_1 = (A_3 + A_4)x_1.$$

Thus our $2N \times 2N$ matrix problem is really an $N \times N$ matrix problem involving $A_1 + A_2$ or $A_3 + A_4$ (it doesn't matter which).

This is precisely the trick we can play with spectral methods in polar coordinates. To be concrete, let us consider the problem of computing the normal modes of oscillation of a circular membrane [MoIn86]. That is, we seek the eigenvalues of the Laplacian on the unit disk:

$$\Delta u = -\lambda^2 u, \quad u = 0 \text{ for } r = 1. \tag{11.4}$$

In polar coordinates the equation takes the form

$$u_{rr} + r^{-1}u_r + r^{-2}u_{\theta\theta} = -\lambda^2 u. \tag{11.5}$$

We can discretize this PDE by a method involving Kronecker products as we have used previously in Programs 16 and 23 (pp. 70 and 93). In (r,θ)-space we have a grid of $(N_r - 1)N_\theta$ points filling the region of the (r,θ) plane indicated in Figure 11.3.

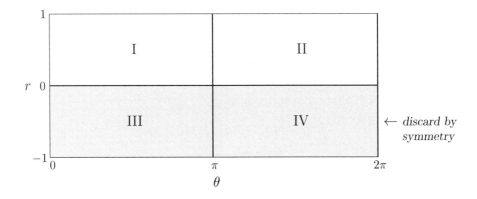

Fig. 11.3. *The map from (r,θ) to (x,y) is 2-to-1, so regions* III *and* IV *in coordinate space can be ignored. Equivalently, one could ignore regions* II *and* IV.

To avoid complications at $r = 0$, we take N_r odd. The discrete Laplacian on the full grid would be an $(N_r - 1)N_\theta \times (N_r - 1)N_\theta$ matrix composed of Kronecker products. However, in view of the symmetry condition (11.2), we will discard the portions of the matrix arising from regions III and IV as redundant. (One could equivalently discard regions II and IV; that is Fornberg's choice.) Still, their effects must be added into the Kronecker products. We do this by dividing our usual matrices for differentiation with respect to r into blocks. Our second derivative in r is a matrix of dimension $(N_r - 1) \times (N_r - 1)$, which we break up as follows:

$$\widetilde{D}_r^2 = \begin{pmatrix} \begin{array}{c|c} D_1 & D_2 \\ \hline D_3 & D_4 \end{array} \end{pmatrix} \begin{array}{l} r>0 \quad \leftarrow\text{added together} \\ r<0 \quad \leftarrow\text{discarded} \end{array}$$

11. Polar Coordinates

Similarly we divide up the first derivative matrix:

$$\widetilde{D}_r = \begin{pmatrix} \overset{r>0}{E_1} & \overset{r<0}{E_2} \\ E_3 & E_4 \end{pmatrix} \begin{matrix} r>0 & \leftarrow\text{added together} \\ \\ r<0 & \leftarrow\text{discarded} \end{matrix}$$

Our second derivative with respect to θ is the matrix $D_\theta^{(2)}$ of (3.12), of dimension $N_\theta \times N_\theta$, and this does not need to be subdivided. All together, following (11.5), our discretization L of the Laplacian in polar coordinates takes the form

$$L = (D_1 + RE_1) \otimes \begin{pmatrix} I & 0 \\ 0 & I \end{pmatrix} + (D_2 + RE_2) \otimes \begin{pmatrix} 0 & I \\ I & 0 \end{pmatrix} + R^2 \otimes D_\theta^{(2)},$$

where I is the $N_{\theta/2} \times N_{\theta/2}$ identity and R is the diagonal matrix

$$R = \text{diag}\left(r_j^{-1}\right), \qquad 1 \leq j \leq (N_r - 1)/2.$$

Program 28 implements this method. The results in Output 28 show four of the eigenmodes of the Laplacian on a disk. The eigenvalues, scaled relative to the lowest one, are accurate to the full precision listed. The eigenmodes are spectrally accurate too; as usual, the roughness of the plot is caused by our displaying the raw grid data rather than the spectral interpolant.

To give a fuller picture of the captivating behavior of this well-known problem, Output 28b plots nodal lines of the first 25 eigenmodes of the same operator. The code that generated this figure was the same as Program 28, except that the `mesh` and `text` commands were deleted, `view(0,20)` was changed to `view(90,0)`, `index = [1 2 6 9]` and `i = 1:4` were changed to `index = 1:25` and `i = 1:25`, and the `subplot` command was changed to a variant of `subplot(5,5,i)` that placed the 25 images close to one another.

Having succeeded in discretizing the Laplacian on the unit disk, we can easily apply it to the solution of other tasks besides calculation of eigenmodes. Following the pattern of Program 16 (p. 70), for example, we can solve the Poisson equation

$$\Delta u = f(r,\theta) = -r^2 \sin(\theta/2)^4 + \sin(6\theta)\cos(\theta/2)^2$$

by solving a linear system of equations. There is no special significance to this right-hand side; it was picked just to make the picture interesting. The solution is computed in Program 29.

Program 28

```
% p28.m - eigenmodes of Laplacian on the disk (compare p22.m)

% r coordinate, ranging from -1 to 1 (N must be odd):
  N = 25; N2 = (N-1)/2;
  [D,r] = cheb(N);
  D2 = D^2;
  D1 = D2(2:N2+1,2:N2+1); D2 = D2(2:N2+1,N:-1:N2+2);
  E1 =  D(2:N2+1,2:N2+1); E2 =  D(2:N2+1,N:-1:N2+2);

% t = theta coordinate, ranging from 0 to 2*pi (M must be even):
  M = 20; dt = 2*pi/M; t = dt*(1:M)'; M2 = M/2;
  D2t = toeplitz([-pi^2/(3*dt^2)-1/6 ...
                 .5*(-1).^(2:M)./sin(dt*(1:M-1)/2).^2]);

% Laplacian in polar coordinates:
  R = diag(1./r(2:N2+1));
  Z = zeros(M2); I = eye(M2);
  L = kron(D1+R*E1,eye(M)) + kron(D2+R*E2,[Z I;I Z]) ...
                          + kron(R^2,D2t);

% Compute four eigenmodes:
  index = [1 3 6 10];
  [V,Lam] = eig(-L); Lam = diag(Lam); [Lam,ii] = sort(Lam);
  ii = ii(index); V = V(:,ii);
  Lam = sqrt(Lam(index)/Lam(1));

% Plot eigenmodes with nodal lines underneath:
  [rr,tt] = meshgrid(r(1:N2+1),[0;t]);
  [xx,yy] = pol2cart(tt,rr);
  z = exp(1i*pi*(-100:100)/100);
  [ay,ax] = meshgrid([.58 .1],[.1 .5]); clf
  for i = 1:4
    u = reshape(real(V(:,i)),M,N2);
    u = [zeros(M+1,1) u([M 1:M],:)];
    u = u/norm(u(:),inf);
    subplot('position',[ax(i) ay(i) .4 .4])
    plot(z), axis(1.05*[-1 1 -1 1 -1 1]), axis off, hold on
    mesh(xx,yy,u)
    view(0,20), colormap([0 0 0]), axis square
    contour3(xx,yy,u-1,[-1 -1])
    plot3(real(z),imag(z),-abs(z))
    text(-.8,4,['Mode ' int2str(index(i))],'fontsize',9)
    text(-.8,3.5, ['\lambda = ', num2str(Lam(i),...
                              '%16.10f')],'fontsize',9)
  end
```

11. Polar Coordinates

Output 28a

Mode 1
$\lambda = 1.0000000000$

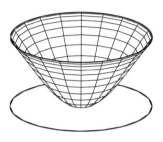

Mode 3
$\lambda = 1.5933405057$

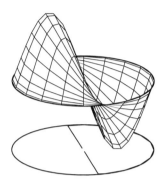

Mode 6
$\lambda = 2.2954172674$

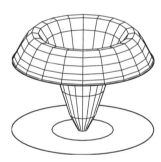

Mode 10
$\lambda = 2.9172954551$

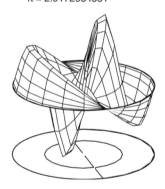

Output 28a: *Eigenmodes of the Laplacian on the unit disk, with nodal curves plotted underneath. Despite the coarse grid, the eigenvalues are accurate to 10 digits.*

Summary of This Chapter. Problems posed in (r, θ) polar coordinates can be solved by spectral methods by using a Chebyshev discretization for r and a Fourier discretization for θ. To weaken the coordinate singularity at $r = 0$, one approach is to take $r \in [-1, 1]$ instead of $r \in [0, 1]$.

Output 28b

Output 28b: *Nodal lines of the first 25 eigenmodes of the Laplacian on a disk. Note that since many of the eigenvalues have multiplicity 2, 11 of the images appear twice with different orientations; only 14 distinct modes are represented.*

Exercises

11.1. What are the first ten eigenvalues of the Laplace operator on the annulus $1 \leq r \leq 2$ with boundary conditions $u = 0$? Find a way to compute them to ten digits of accuracy by a two-dimensional spectral method. Then use separation of variables to reduce the problem to one dimension, and solve it again either numerically or analytically.

11. Polar Coordinates

Program 29

```
% p29.m - solve Poisson equation on the unit disk
%         (compare p16.m and p28.m)

% Laplacian in polar coordinates:
  N = 31; [D,r] = cheb(N); N2 = (N-1)/2; D2 = D^2;
  D1 = D2(2:N2+1,2:N2+1); D2 = D2(2:N2+1,N:-1:N2+2);
  E1 =  D(2:N2+1,2:N2+1); E2 =  D(2:N2+1,N:-1:N2+2);
  M = 40; dt = 2*pi/M; t = dt*(1:M)'; M2 = M/2;
  D2t = toeplitz([-pi^2/(3*dt^2)-1/6 ...
                  .5*(-1).^(2:M)./sin(dt*(1:M-1)/2).^2]);
  R = diag(1./r(2:N2+1)); Z = zeros(M2); I = eye(M2);
  L = kron(D1+R*E1,eye(M))+kron(D2+R*E2,[Z I;I Z])+kron(R^2,D2t);

% Right-hand side and solution for u:
  [rr,tt] = meshgrid(r(2:N2+1),t); rr = rr(:); tt = tt(:);
  f = -rr.^2.*sin(tt/2).^4 + sin(6*tt).*cos(tt/2).^2; u = L\f;

% Reshape results onto 2D grid and plot them:
  u = reshape(u,M,N2); u = [zeros(M+1,1) u([M 1:M],:)];
  [rr,tt] = meshgrid(r(1:N2+1),t([M 1:M]));
  [xx,yy] = pol2cart(tt,rr);
  clf, subplot('position',[.1 .4 .8 .5])
  mesh(xx,yy,u), view(20,40), colormap([0 0 0])
  axis([-1 1 -1 1 -.01 .05]), xlabel x, ylabel y, zlabel u
```

Output 29

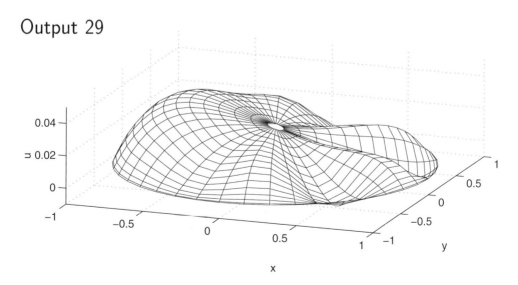

Output 29: *Poisson equation on the unit disk.*

11.2. Output 28a looks choppy because the raw data on the spectral grid are plotted rather than the implicit spectral interpolant. Find a way to modify Program 28 so as to plot the latter, and produce a corresponding figure analogous to Output 28a.

11.3. Suppose the PDE of (11.4) is modified to $\Delta u = -\lambda^2(1 + x/2)u$, so that the coefficient varies from one side of the circular membrane to the other. Determine to high accuracy the new lowest three eigenvalues, and plot the corresponding eigenmodes.

11.4. It was mentioned on p. 55 and in Exercise 6.4 that Chebyshev differentiation matrices have the symmetry property $(D_N)_{ij} = -(D_N)_{N-i,N-j}$. Taking N odd, for simplicity, so that D_N is of even dimension, show how, by decomposing D_N in a fashion analogous to the decomposition of (11.3), one can compute matrix-vector products $D_N v$ via a matrix E that has only half the size of D_N. In [Sol92] it is shown that this trick can speed up certain computations by a factor of 2.

12. Integrals and Quadrature Formulas

Up to now we have been solving ODEs, PDEs, and related eigenvalue problems. Now suppose that we are faced with the simpler task of evaluating an integral such as

$$I = \int_{-1}^{1} f(x)\,dx. \tag{12.1}$$

How could we compute I by a spectral method?

One approach is to note that an integral is the special case of an ODE $u' = f(x, u)$ in which f is independent of u. Thus (12.1) can be restated as the initial value problem

$$u'(x) = f(x), \quad u(-1) = 0, \quad x > -1, \tag{12.2}$$

where our goal is to evaluate $I = u(1)$. For this we can set up a spectral method on $[-1, 1]$ on our usual Chebyshev grid. To impose the boundary condition $u(-1) = 0$, we strip off the last row and column of the differentiation matrix D_N in the usual manner described in Chapter 7. If $\widetilde{D}_N$ is the resulting matrix of dimension $N \times N$, we are left with the linear system of equations

$$\widetilde{D}_N v = f$$

with $f = (f(x_0), \ldots, f(x_{N-1}))^T$. Our approximation to I is given by $I_N = v_0$.

In fact, since we care only about the first component of v, there is no need to solve the whole system of equations. If we let w^T denote the first row of $\widetilde{D}_N^{-1}$, a row vector of length N, then another formula for the same approximation is

$$I_N = w^T f. \qquad (12.3)$$

Speaking abstractly, integration over $[-1, 1]$ is a linear functional $I(f)$, and a linear numerical approximation to I based on discrete data will constitute another linear functional $I_N(f)$. Equation (12.3) expresses the fact that any linear functional is equivalent to an inner product with some weight vector w—the *Riesz representation theorem* [LiLo97].

Program 30, whose output is labeled Output 30a, illustrates the success of this method by integrating $|x|^3$, $\exp(-x^{-2})$, $1/(1+x^2)$, and x^{10}, the same functions that we examined in Chapter 6 in connection with the convergence of spectral differentiation. Spectral accuracy just as in Output 12 (p. 58) is evident. Note that x^{10} is integrated exactly for $N \geq 11$.

The method just described made use of our existing spectral differentiation matrix. An alternative and better approach is to start from our fundamental spectral philosophy:

- *Find the polynomial of degree $\leq N$ such that $p(x_j) = f_j$, $0 \leq j \leq N$.*
- *Set $I_N = \int_{-1}^{1} p(x)\,dx$.*

This formulation must be different, for it will integrate x^{10} exactly for $N \geq 10$ rather than $N \geq 11$. (It makes use of the value $f(-1)$, which the previous method ignored.) In fact, this new strategy goes by the name of *Clenshaw–Curtis quadrature* [ClCu60]. In the field of numerical integration [DaRa84, KrUe98], it can be classed as the formula of optimal order based on the fixed set of Chebyshev nodes $\{x_j\}$—as opposed to the Gauss formula of optimal order based on optimally chosen nodes, which we shall discuss in a moment.

One way to compute the Clenshaw–Curtis approximation would be by using the FFT methods of Chapter 8. Given a function $f(x)$ defined on $[-1, 1]$, consider the self-reciprocal function $\mathsf{f}(z)$ defined on the unit circle $|z|=1$ by the 2-to-1 pointwise equivalence $x = \mathrm{Re}\, z$ of (8.1). If $p(x) = \sum_{n=0}^{N} a_n T_n(x)$ is the polynomial interpolant to $f(x)$ in the Chebyshev points $\{x_j\}$, then $p(x)$ corresponds pointwise to the self-reciprocal Laurent polynomial interpolant $\mathsf{p}(z) = \frac{1}{2}\sum_{n=0}^{N} a_n(z^n + z^{-n})$ to $\mathsf{f}(z)$ in roots of unity $\{z_j\}$. Since $x = \frac{1}{2}(z+z^{-1})$ and $dx/dz = \frac{1}{2}(1 - z^{-2})$, we compute

$$\begin{aligned}
\int_{-1}^{1} p(x)\,dx &= \int_{-1}^{1} \mathsf{p}(z)\,dz\,\frac{dx}{dz} \\
&= \frac{1}{4}\sum_{n=0}^{N} a_n \int_{-1}^{1} (z^n + z^{-n})(1 - z^{-2})\,dz
\end{aligned}$$

12. Integrals and Quadrature Formulas

Program 30

```
% p30.m - spectral integration, ODE style (compare p12.m)
% Computation: various values of N, four functions:
  Nmax = 50; E = zeros(4,Nmax); clf
  for N = 1:Nmax; i = 1:N;
    [D,x] = cheb(N); x = x(i); Di = inv(D(i,i)); w = Di(1,:);
    f = abs(x).^3;      E(1,N) = abs(w*f - .5);
    f = exp(-x.^(-2));  E(2,N) = abs(w*f - ...
                        2*(exp(-1)+sqrt(pi)*(erf(1)-1)));
    f = 1./(1+x.^2);    E(3,N) = abs(w*f - pi/2);
    f = x.^10;          E(4,N) = abs(w*f - 2/11);
  end

% Plot results:
  labels = {'|x|^3','exp(-x^{-2})','1/(1+x^2)','x^{10}'};
  for iplot = 1:4,
    subplot(3,2,iplot)
    semilogy(E(iplot,:)+1e-100,'.','markersize',12), hold on
    plot(E(iplot,:)+1e-100)
    axis([0 Nmax 1e-18 1e3]), grid on
    set(gca,'xtick',0:10:Nmax,'ytick',(10).^(-15:5:0))
    ylabel error, text(32,.004,labels(iplot))
  end
```

Output 30a

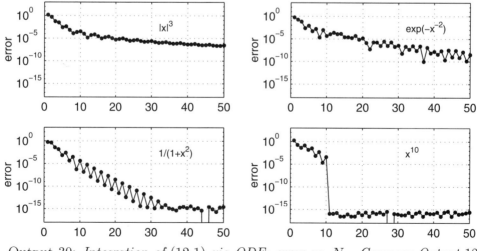

Output 30: *Integration of* (12.1) *via ODE: error vs.* N. *Compare Output* 12 *(p. 58)*.

$$= \frac{1}{4}\sum_{n=0}^{N} a_n \int_{-1}^{1} \left(z^n - z^{n-2} + z^{-n} - z^{-n-2}\right)$$

$$= \frac{1}{4}\sum_{n=0}^{N} a_n \left(\left.\frac{z^{n+1} + z^{-n-1}}{n+1}\right|_{-1}^{1} - \left.\frac{z^{n-1} + z^{-n+1}}{n-1}\right|_{-1}^{1}\right)$$

$$= \sum_{\substack{n=0 \\ n \text{ even}}}^{N} a_n \left(\frac{1}{n+1} - \frac{1}{n-1}\right) = \sum_{\substack{n=0 \\ n \text{ even}}}^{N} \frac{2a_n}{1-n^2}.$$

Thus to implement Clenshaw–Curtis quadrature, we can use the FFT to determine the coefficients $\{a_n\}$ as in Chapter 8, then sum the results over even values of n with the weights $2/(1-n^2)$.

This method works, but it is more elaborate than necessary, for by pursuing the algebra a little further, one can determine the Clenshaw–Curtis weights analytically. Rather than write down the results in formulas, we encapsulate them in a MATLAB program:

clencurt.m

```
% CLENCURT  nodes x (Chebyshev points) and weights w
%           for Clenshaw-Curtis quadrature
  function [x,w] = clencurt(N)
  theta = pi*(0:N)'/N; x = cos(theta);
  w = zeros(1,N+1); ii = 2:N; v = ones(N-1,1);
  if mod(N,2)==0
    w(1) = 1/(N^2-1); w(N+1) = w(1);
    for k=1:N/2-1, v = v - 2*cos(2*k*theta(ii))/(4*k^2-1); end
    v = v - cos(N*theta(ii))/(N^2-1);;
  else
    w(1) = 1/N^2; w(N+1) = w(1);
    for k=1:(N-1)/2, v = v - 2*cos(2*k*theta(ii))/(4*k^2-1); end
  end
  w(ii) = 2*v/N;
```

Output 30b shows the results obtained by modifying Program 30 to use clencurt. They are marginally more accurate than before, and much cleaner.

The convergence rates exhibited in Outputs 30a and 30b are excellent—this is spectral accuracy of the kind showcased throughout this book. Nevertheless, we can do better. If we use a Gaussian formula, then the integral will be exact for polynomials of degree $2N - 1$, not just N or $N - 1$. For this we must

12. Integrals and Quadrature Formulas

Output 30b

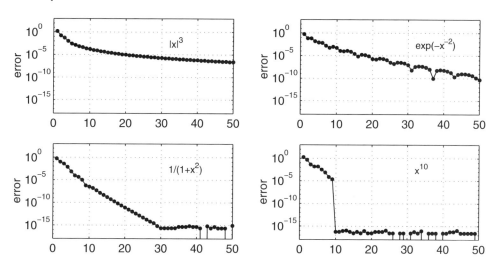

Output 30b: *Clenshaw–Curtis integration of* (12.1). *These results are generated by Program 30 except with the line beginning* `[D,x] = ...` *replaced by the command* `[x,w] = clencurt(N)`.

take $\{x_j\}$ to be not Chebyshev points but *Legendre points*, that is, roots of Legendre polynomials in $(-1, 1)$. These points and the associated weights can be computed numerically by solving a tridiagonal matrix eigenvalue problem [GoWe69, TrBa97]. The next, surprisingly short program specifies the details.

gauss.m

```
% GAUSS   nodes x (Legendre points) and weights w
%         for Gauss quadrature

  function [x,w] = gauss(N)
  beta = .5./sqrt(1-(2*(1:N-1)).^(-2));
  T = diag(beta,1) + diag(beta,-1);
  [V,D] = eig(T);
  x = diag(D); [x,i] = sort(x);
  w = 2*V(1,i).^2;
```

Output 30c shows the results obtained with Gauss quadrature. Note that for the smoother functions, the convergence surpasses that of Outputs 30a and 30b, but there is not much difference for the functions that are less smooth.

Output 30c

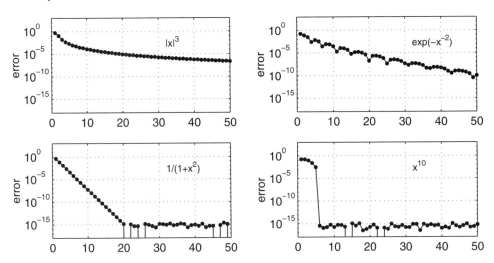

Output 30c: *Gauss integration of* (12.1). *Here we run Program 30 again, but with the command* [x,w] = gauss(N). *Note that the convergence is now faster for the smoother functions f.*

Gauss quadrature has genuine advantages over Clenshaw–Curtis quadrature for definite integrals. However, most applications of spectral methods involve the solution of differential equations. For these problems, Gauss quadrature is still relevant if one solves the problem by a Galerkin formulation, but it is less relevant for solutions by collocation, as in this book. Some practitioners feel strongly that Galerkin formulations are superior; others feel they require extra effort for little gain. For better or worse, the present book concentrates on collocation, and we shall make no further use of Gauss quadrature.

All of the discussion in this chapter has considered integration by Chebyshev spectral methods and their variants, not Fourier methods. What about the latter? Are there problems where we wish to calculate integrals over periodic domains, and do Fourier spectral methods provide a useful technique for such problems?

The answer is smashingly yes. Suppose we wish to evaluate an integral

$$I = \int_0^{2\pi} f(\theta)\, d\theta, \qquad (12.4)$$

where f is 2π-periodic. According to the usual spectral collocation philosophy, we will construct a trigonometric interpolant in equispaced points and then integrate the interpolant. In this integral, all the nonconstant terms will integrate to zero, leaving us with just the constant term. That is, periodic

12. Integrals and Quadrature Formulas

Fourier integration reduces to the *periodic trapezoid rule*,

$$I_N = \frac{2\pi}{N} \sum_{j=1}^{N} f(\theta_j), \qquad (12.5)$$

with $\theta_j = j\pi/N$ as usual. Our weight vector w is a multiple of $(1,1,1,\ldots,1)^T$.

For smooth integrands, for the usual reasons analyzed in Chapter 4, it follows that the periodic trapezoid rule converges extraordinarily fast. For illustration, suppose we use (12.5) to determine the perimeter of an ellipse of length 2 and width 1, which is given by the integral

$$\int_0^{2\pi} \left(\tfrac{1}{4}\sin^2\theta + \cos^2\theta\right)^{1/2} d\theta.$$

The single line of MATLAB

```
t=2*pi*(1:N)/N; I=2*pi*mean(sqrt(.25*sin(t).^2+cos(t).^2))
```

is enough to carry out this computation, and with $N = 25$, we get $I_N = 4.84422411027386$, which is correct except in the last digit. (The number in question is $4E(3/4)$, where E is the complete elliptic integral of the second kind [AbSt65]; in MATLAB, `[K,E] = ellipke(3/4)`, `perimeter = 4*E`.) For more on this phenomenon of rapid convergence of the periodic trapezoid rule, see [DaRa84], [Hen86], and Exercise 12.6.

There is a special context in which integrals over periodic domains regularly arise: as contour integrals in the complex plane. This is a beautiful subject which, although off the beaten track of spectral methods, is a standard tool in computational complex analysis. If $f(z)$ is an analytic function in the closed unit disk, for example, then its Taylor series converges there, and the Taylor coefficients can be computed by Cauchy integrals:

$$f(z) = \sum_{j=0}^{\infty} a_j z^j, \qquad a_j = \frac{1}{2\pi i} \int_{|z|=1} z^{-1-j} f(z)\, dz, \qquad (12.6)$$

where the contour of integration is the unit circle traversed once counterclockwise. (If $f(z)$ is merely analytic in a neighborhood of the unit circle, not throughout the disk, the formulas generalize to a *Laurent series*, convergent in an annulus, with terms $-\infty < j < \infty$.) Setting $z = e^{i\theta}$, with $dz = iz d\theta$, shows that an equivalent expression for a_j is

$$a_j = \frac{1}{2\pi} \int_0^{2\pi} e^{-ij\theta} f(e^{i\theta})\, d\theta. \qquad (12.7)$$

Thus each coefficient of a Taylor series can be evaluated accurately by the periodic trapezoid rule. What is more remarkable is that a whole collection of coefficients can be evaluated simultaneously by the FFT (Exercise 12.7). This observation forms the basis of fast algorithms for problems in computational complex analysis as diverse as differentiation, integration, analytic continuation, zerofinding, computation of transforms, evaluation of special functions, and conformal mapping [Hen79, Hen86].

Here is an example involving just one trapezoid rule integral, not the FFT. One of the most familiar of special functions is the gamma function $\Gamma(z)$, the complex generalization of the factorial function, which satisfies $\Gamma(n+1) = n!$ for each integer $n \geq 0$. $\Gamma(z)$ has a pole at each of the nonpositive integers, but $1/\Gamma(z)$ is analytic for all z and is given by a contour integral formula due to Hankel (see equation (8.8.23) of [Hil62]),

$$\frac{1}{\Gamma(z)} = \frac{1}{2\pi i} \int_C e^t t^{-z} \, dt, \qquad (12.8)$$

where C is a contour in the complex plane that begins at $-\infty - 0i$ (just below the branch cut of t^{-z} on the negative real axis), winds counterclockwise once around the origin, and ends at $-\infty + 0i$ (just above). Since the integrand decays exponentially as $\operatorname{Re} t \to -\infty$, we can get results as accurate as we like by replacing C by a bounded contour that begins and ends sufficiently far out on the negative real axis. Specifically, Program 31 takes C to be the circle of radius $r = 16$ centered at $c = -11$. If we define $t = c + re^{i\theta}$, then we have $dt = ire^{i\theta} = i(t-c)$, and the integral becomes

$$\frac{1}{\Gamma(z)} = \frac{1}{2\pi} \int_{-\pi}^{\pi} e^t t^{-z}(t-c) \, d\theta. \qquad (12.9)$$

If we evaluate this by the trapezoid rule (12.5), we find that $1/\Gamma(z)$ is approximated simply by the mean value of $e^t t^{-z}(t-c)$ over equispaced points on the contour C. It couldn't be much simpler! Output 31 inverts the result to show the familiar shape of the gamma function generated to high accuracy by this technique.

Summary of This Chapter. The natural spectral method for numerical integration in Chebyshev points is Clenshaw–Curtis quadrature, defined by integrating the polynomial interpolant, and it is spectrally accurate. A higher order of spectral accuracy can be achieved by Gauss quadrature, based on interpolation in Legendre points instead, and this is the basis of many Galerkin spectral methods. The natural spectral integration formula on a periodic interval or a closed contour in the complex plane is the trapezoid rule, and in conjunction with the FFT, this has powerful applications in complex analysis.

12. Integrals and Quadrature Formulas

Program 31

```
% p31.m - gamma function via complex integral, trapezoid rule
  N = 70; theta = -pi + (2*pi/N)*(.5:N-.5)';
  c = -11;                      % center of circle of integration
  r = 16;                       % radius of circle of integration
  x = -3.5:.1:4; y = -2.5:.1:2.5;
  [xx,yy] = meshgrid(x,y); zz = xx + 1i*yy; gaminv = 0*zz;
  for i = 1:N
    t = c + r*exp(1i*theta(i));
    gaminv = gaminv + exp(t)*t.^(-zz)*(t-c);
  end
  gaminv = gaminv/N; gam = 1./gaminv; clf, mesh(xx,yy,abs(gam))
  axis([-3.5 4 -2.5 2.5 0 6]), xlabel Re(z), ylabel Im(z)
  text(4,-1.4,5.5,'|\Gamma(z)|','fontsize',20), colormap([0 0 0])
```

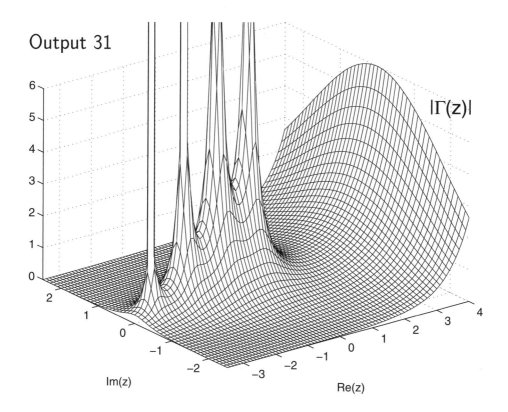

Output 31: *Computation of the gamma function by a 70-point trapezoid rule approximation to the contour integral (12.8). At most points of the grid, the computed result is accurate to 8 digits.*

Exercises

12.1. Perform a comparative study of Chebyshev vs. Legendre points. To make the comparisons as close as possible, define Chebyshev points via zeros rather than extrema as in (6.1): $x_j = \cos((j-1/2)\pi/N)$, $j = 1, 2, \ldots, N$. Plot the two sets of points for $N = 5, 10, 15$, and find a graphical way to compare their locations as $N \to \infty$. Modify Programs 9 and 10 to use Legendre instead of Chebyshev points, and discuss how the results compare with those of Outputs 9 and 10.

12.2. Write a MATLAB program to implement (6.8) and (6.9) and construct the differentiation matrix D_N associated with an arbitrary set of distinct points $x_0, \ldots, x_N$. Combine it with `gauss` to create a function that computes the matrix D_N associated with Legendre points in $(-1, 1)$. Print results for $N = 1, 2, 3, 4$.

12.3. Suppose you didn't know about Clenshaw–Curtis quadrature and had to reinvent it. One approach would be to find the weights by setting up and solving an appropriate system of linear equations in Vandermonde form. Describe the mathematics of this process, and then implement it with the help of MATLAB's command `vander`. Compare the weight vectors w obtained in this manner with those delivered by `clencurt` for $N = 4, 8$, and 128.

12.4. Write a program based on a Chebyshev spectral method to compute the indefinite integral $f(x) = \int_0^x \sin(6s^{2.5}) ds$ for $0 \leq x \leq 2$. The program should plot values at (shifted) Chebyshev points and the curve of the polynomial interpolant between these values, and print the error $f(1)_{\text{computed}} - f(1)_{\text{exact}}$. Produce results for $N = 10, 20, 30, 40, 50$. Comment on the accuracy as a function of N and on how the accuracy appears to depend on the local number of points per wavelength.

12.5. To 10 digits, what is the perimeter of the superellipse defined by the equation $x^4 + y^4 = 1$? To 10 digits, what exponent α has the property that the curve defined by the equation $|x|^\alpha + |y|^\alpha = 1$ has perimeter equal to 7?

12.6. Suppose the 2π-periodic function $f(x)$ extends to an analytic function in the strip $|\text{Im}(z)| < a$ in the complex plane for some $a > 0$. From results of Chapter 4, derive an estimate for the error in evaluating $\int_{-\pi}^{\pi} f(x) dx$ by the trapezoid rule with step size h. Perform the integration numerically for the function $f(x) = (1 + \sin^2(x/2))^{-1}$ of Program 7 (p. 35). Does the actual convergence behavior match your estimate?

12.7. Use the FFT in N points to calculate the first 20 Taylor series coefficients of $f(z) = \log(1 + \frac{1}{2}z)$. What is the asymptotic convergence factor as $N \to \infty$? Can you explain this number?

12.8. What symmetry property does $1/\Gamma(z)$ satisfy with respect to the real axis? When c is real as in Program 31, the computed estimates of $1/\Gamma(z)$ will satisfy the same symmetry property. If c is moved off the real axis, however, the magnitude of the resulting loss of symmetry can be used to give some idea of the error in the computation. Try this with $c = -11 + i$ and produce a contour plot of the error estimate with contours at $10^{-5}, 10^{-6}, 10^{-7}, \ldots$. How does your contour plot change if N is increased to 100?

13. More about Boundary Conditions

So far we have treated just simple homogeneous Dirichlet boundary conditions $u(\pm 1) = 0$, as well as periodic boundary conditions. Of course, many problems require more than this, and in this chapter we outline some of the techniques available.

There are two basic approaches to boundary conditions for spectral collocation methods:

(I) *Restrict attention to interpolants that satisfy the boundary conditions;* or

(II) *Do not restrict the interpolants, but add additional equations to enforce the boundary conditions.*

So far we have only used method (I), but method (II) is more flexible and is often better for more complicated problems. (It is related to the so-called *tau methods* that appear in the field of Galerkin spectral methods.)

We begin with another example involving method (I). In Program 13 (p. 64) we solved $u_{xx} = e^{4x}$ on $[-1, 1]$ subject to $u(-1) = u(1) = 0$. Consider now instead the inhomogeneous problem

$$u_{xx} = e^{4x}, \quad -1 < x < 1, \ u(-1) = 0, \ u(1) = 1. \tag{13.1}$$

Method (I) can be applied in this case too, with embarrassing ease. Since the equation is linear and the second derivative of x is zero, we can simply solve the problem with $u(\pm 1) = 0$ and then add $(x+1)/2$ to the result. See Program 32.

Program 32

```
% p32.m - solve u_xx = exp(4x), u(-1)=0, u(1)=1 (compare p13.m)
  N = 16;
  [D,x] = cheb(N);
  D2 = D^2;
  D2 = D2(2:N,2:N);
  f = exp(4*x(2:N));
  u = D2\f;
  u = [0;u;0] + (x+1)/2;
  clf
  subplot('position',[.1 .4 .8 .5])
  plot(x,u,'.','markersize',16)
  xx = -1:.01:1;
  uu = polyval(polyfit(x,u,N),xx);
  line(xx,uu), grid on
  exact = (exp(4*xx) - sinh(4)*xx - cosh(4))/16 + (xx+1)/2;
  title(['max err = ' num2str(norm(uu-exact,inf))],'fontsize',12)
```

Output 32

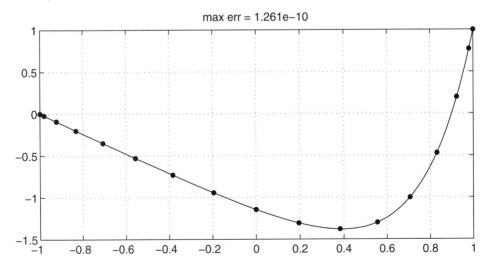

Output 32: *Solution of the boundary value problem* (13.1) *with inhomogeneous boundary data.*

13. More about Boundary Conditions

Now, suppose we are faced with the same ODE with a Neumann condition at the left endpoint,

$$u_{xx} = e^{4x}, \quad -1 < x < 1, \quad u_x(-1) = u(1) = 0. \tag{13.2}$$

This time, it is convenient to turn to method (II). At $x = 1$, i.e., grid point $j = 0$, we will delete a row and a column of the differentiation matrix as usual. At $x = -1$ and $j = N$, on the other hand, we wish to impose a condition involving the first derivative. What could be more natural than to use the spectral differentiation matrix D for this purpose? Thus we will end up solving an $N \times N$ (not $(N-1) \times (N-1)$) linear system of equations in which the first N equations enforce the condition $u_{xx} = e^{4x}$ at the interior grid points and the final equation enforces the condition $u_x = 0$ at the leftmost grid point. The matrix of the system of equations will contain $N-1$ rows extracted from $(D_N)^2$ and one taken from D_N. The details appear in Program 33, and in Output 33, we see that nine-digit accuracy is achieved with $N = 16$.

The use of similar methods for a more interesting equation is illustrated in Program 34. The *Allen–Cahn* or *bistable equation* is an example of a nonlinear *reaction-diffusion equation*:

$$u_t = \epsilon u_{xx} + u - u^3, \tag{13.3}$$

where ϵ is a parameter. This equation has three constant steady states, $u = -1$, $u = 0$, and $u = 1$. The middle state is unstable, but the states $u = \pm 1$ are attracting, and solutions tend to exhibit flat areas close to these values separated by interfaces that may coalesce or vanish on a long time scale, a phenomenon known as *metastability*. In Output 34 we see metastability up to $t \approx 45$ followed by rapid transition to a solution with just one interface.

Now, what if we had more complicated boundary conditions, such as

$$u(-1, t) = 0, \quad u(1, t) = 1 + \sin^2(t/5)? \tag{13.4}$$

Here it again becomes convenient to switch to method (II), and Program 35 illustrates how this can be done. Since $1 + \sin^2(t/5) > 1$ for most t, the boundary condition effectively pumps amplitude into the system, and the effect is that the location of the final interface is moved from $x = 0$ to $x \approx -0.4$. Notice also that the transients vanish earlier, at $t \approx 30$ instead of $t \approx 45$.

Program 36 illustrates the same kind of methods for a time-independent problem, the Laplace equation

$$u_{xx} + u_{yy} = 0, \quad -1 < x, y < 1, \tag{13.5}$$

subject to the boundary conditions

$$u(x, y) = \begin{cases} \sin^4(\pi x), & y = 1 \text{ and } -1 < x < 0, \\ \frac{1}{5} \sin(3\pi y), & x = 1, \\ 0, & \text{otherwise.} \end{cases} \tag{13.6}$$

Program 33

```
% p33.m - solve linear BVP u_xx = exp(4x), u'(-1)=u(1)=0

  N = 16;
  [D,x] = cheb(N); D2 = D^2;
  D2(N+1,:) = D(N+1,:);                  % Neumann condition at x = -1
  D2 = D2(2:N+1,2:N+1);
  f = exp(4*x(2:N));
  u = D2\[f;0];
  u = [0;u];
  clf, subplot('position',[.1 .4 .8 .5])
  plot(x,u,'.','markersize',16)
  axis([-1 1 -4 0])
  xx = -1:.01:1;
  uu = polyval(polyfit(x,u,N),xx);
  line(xx,uu)
  grid on
  exact = (exp(4*xx) - 4*exp(-4)*(xx-1) - exp(4))/16;
  title(['max err = ' num2str(norm(uu-exact,inf))],'fontsize',12)
```

Output 33

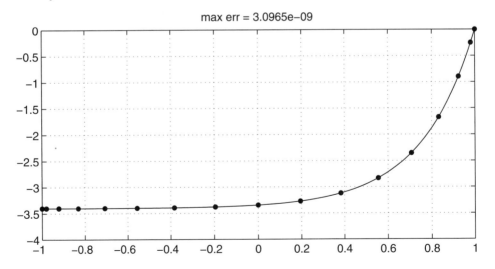

Output 33: *Solution of the boundary value problem (13.2) with a Neumann boundary condition.*

13. More about Boundary Conditions

Program 34

```
% p34.m - Allen-Cahn eq. u_t = u_xx + u - u^3, u(-1)=-1, u(1)=1
%          (compare p6.m and p32.m)

% Differentiation matrix and initial data:
  N = 20; [D,x] = cheb(N); D2 = D^2;       % use full-size matrix
  D2([1 N+1],:) = zeros(2,N+1);            % for convenience
  eps = 0.01; dt = min([.01,50*N^(-4)/eps]);
  t = 0; v = .53*x + .47*sin(-1.5*pi*x);

% Solve PDE by Euler formula and plot results:
  tmax = 100; tplot = 2; nplots = round(tmax/tplot);
  plotgap = round(tplot/dt); dt = tplot/plotgap;
  xx = -1:.025:1; vv = polyval(polyfit(x,v,N),xx);
  plotdata = [vv; zeros(nplots,length(xx))]; tdata = t;
  for i = 1:nplots
    for n = 1:plotgap
      t = t+dt; v = v + dt*(eps*D2*(v-x) + v - v.^3);   % Euler
    end
    vv = polyval(polyfit(x,v,N),xx);
    plotdata(i+1,:) = vv; tdata = [tdata; t];
  end
  clf, subplot('position',[.1 .4 .8 .5])
  mesh(xx,tdata,plotdata), grid on, axis([-1 1 0 tmax -1 1]),
  view(-60,55), colormap([0 0 0]), xlabel x, ylabel t, zlabel u
```

Output 34

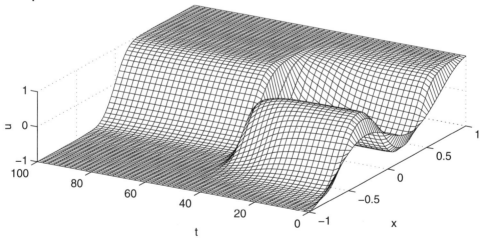

Output 34: *Solution of the Allen–Cahn equation* (13.3) *with* $\epsilon = 10^{-2}$. *The interior humps are metastable and vanish suddenly near* $t = 45$.

Method (II) is used to enforce the boundary conditions; the mathematics is straightforward but one must be careful with the bookkeeping. MATLAB supports cleverness in such matters well with its logical operations and commands such as "`find`," and if the reader understands Program 36 in detail, he or she is ready for MATLAB expert slopes!

Program 37 gives another example of the use of Neumann boundary conditions. Following Program 20 (pp. 83–84), we again consider the second-order wave equation in two dimensions, now on a rectangular domain:

$$u_{tt} = u_{xx} + u_{yy}, \qquad -3 < x < 3, \quad -1 < y < 1. \tag{13.7}$$

What is new are Neumann boundary conditions along the sides of this "wave tank" and periodic boundary conditions at the ends:

$$u_y(x, \pm 1, t) = 0, \qquad u(-3, y, t) = u(3, y, t). \tag{13.8}$$

Program 37 approximates this problem by Fourier discretization in x, Chebyshev discretization in y, and the leap frog formula in t, and the Neumann boundary data are imposed essentially as in Program 33. The initial conditions are chosen for programming convenience:

$$u(x, y, 0) = e^{-8((x+\frac{3}{2})^2 + y^2)}, \quad u(x, y, -\Delta t) = e^{-8((x+\Delta t + \frac{3}{2})^2 + y^2)}.$$

If the Gaussian pulse of $u(x, y, 0)$ traveling rightward at speed 1 were a solution of (13.7), then this choice of $u(x, y, -\Delta t)$ would amount to the imposition of initial conditions appropriate to that solution; but it is not, and the initial conditions approximated by Program 37 can best be written as simply

$$u(x, y, 0) = e^{-8((x+\frac{3}{2})^2 + y^2)}, \quad u_t(x, y, 0) = -u_x(x, y, 0). \tag{13.9}$$

Summary of This Chapter. Simple boundary conditions for spectral collocation methods can be imposed by restricting attention to interpolants that satisfy the boundary conditions. For more complicated problems, it is more convenient to permit arbitrary interpolants but add additional equations to the discrete problem to enforce the boundary conditions.

Exercises

13.1. Suppose (as in Program 35) that a time-dependent initial boundary value problem $u_t = L(u)$ is posed on $[a, b]$ subject to time-dependent Dirichlet boundary conditions $u(a, t) = u_a(t)$, $u(b, t) = u_b(t)$. If $w(x, t)$ is an arbitrary smooth function that satisfies the boundary conditions, show how w can be used to reduce the problem to one with homogeneous boundary conditions.

13. More about Boundary Conditions

Program 35

```
% p35.m - Allen-Cahn eq. as in p34.m, but with boundary condition
%           imposed explicitly ("method (II)")

% Differentiation matrix and initial data:
  N = 20; [D,x] = cheb(N); D2 = D^2;
  eps = 0.01; dt = min([.01,50*N^(-4)/eps]);
  t = 0; v = .53*x + .47*sin(-1.5*pi*x);

% Solve PDE by Euler formula and plot results:
  tmax = 100; tplot = 2; nplots = round(tmax/tplot);
  plotgap = round(tplot/dt); dt = tplot/plotgap;
  xx = -1:.025:1; vv = polyval(polyfit(x,v,N),xx);
  plotdata = [vv; zeros(nplots,length(xx))]; tdata = t;
  for i = 1:nplots
    for n = 1:plotgap
      t = t+dt; v = v + dt*(eps*D2*v + v - v.^3);      % Euler
      v(1) = 1 + sin(t/5)^2;  v(end) = -1;             % BC
    end
    vv = polyval(polyfit(x,v,N),xx);
    plotdata(i+1,:) = vv; tdata = [tdata; t];
  end
  clf, subplot('position',[.1 .4 .8 .5])
  mesh(xx,tdata,plotdata), grid on, axis([-1 1 0 tmax -1 2]),
  view(-60,55), colormap([0 0 0]), xlabel x, ylabel t, zlabel u
```

Output 35

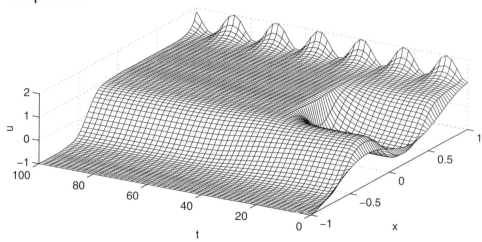

Output 35: *Same as Output 34, but with a nonconstant boundary condition at $x = 1$ imposed by method* (II).

Program 36

```
% p36.m - Laplace eq. on [-1,1]x[-1,1] with nonzero BCs

% Set up grid and 2D Laplacian, boundary points included:
  N = 24; [D,x] = cheb(N); y = x;
  [xx,yy] = meshgrid(x,y); xx = xx(:); yy = yy(:);
  D2 = D^2; I = eye(N+1); L = kron(I,D2) + kron(D2,I);

% Impose boundary conditions by replacing appropriate rows of L:
  b = find(abs(xx)==1 | abs(yy)==1);           % boundary pts
  L(b,:) = zeros(4*N,(N+1)^2); L(b,b) = eye(4*N);
  rhs = zeros((N+1)^2,1);
  rhs(b) = (yy(b)==1).*(xx(b)<0).*sin(pi*xx(b)).^4 + ...
           .2*(xx(b)==1).*sin(3*pi*yy(b));

% Solve Laplace equation, reshape to 2D, and plot:
  u = L\rhs; uu = reshape(u,N+1,N+1);
  [xx,yy] = meshgrid(x,y);
  [xxx,yyy] = meshgrid(-1:.04:1,-1:.04:1);
  uuu = interp2(xx,yy,uu,xxx,yyy,'cubic');
  clf, subplot('position',[.1 .4 .8 .5])
  mesh(xxx,yyy,uuu), colormap([0 0 0])
  axis([-1 1 -1 1 -.2 1]), view(-20,45)
  text(0,.8,.4,sprintf('u(0,0) = %12.10f',uu(N/2+1,N/2+1)))
```

Output 36

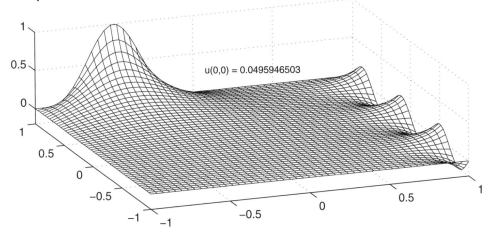

Output 36: *Laplace equation with nonzero boundary data.*

13. More about Boundary Conditions

Program 37

```
% p37.m - 2D "wave tank" with Neumann BCs for |y|=1

% x variable in [-A,A], Fourier:
  A = 3; Nx = 50; dx = 2*A/Nx; x = -A+dx*(1:Nx)';
  D2x = (pi/A)^2*toeplitz([-1/(3*(dx/A)^2)-1/6 ...
          .5*(-1).^(2:Nx)./sin((pi*dx/A)*(1:Nx-1)/2).^2]);

% y variable in [-1,1], Chebyshev:
  Ny = 15; [Dy,y] = cheb(Ny); D2y = Dy^2;
  BC = -Dy([1 Ny+1],[1 Ny+1])\Dy([1 Ny+1],2:Ny);

% Grid and initial data:
  [xx,yy] = meshgrid(x,y);
  vv = exp(-8*((xx+1.5).^2+yy.^2));
  vvold = exp(-8*((xx+dt+1.5).^2+yy.^2));

% Time-stepping by leap frog formula:
  dt = 5/(Nx+Ny^2); clf, plotgap = round(2/dt); dt = 2/plotgap;
  for n = 0:2*plotgap
    t = n*dt;
    if rem(n+.5,plotgap)<1
      subplot(3,1,n/plotgap+1), mesh(xx,yy,vv), view(-10,60)
      axis([-A A -1 1 -0.15 1]), colormap([0 0 0])
      text(-2.5,1,.5,['t = ' num2str(t)],'fontsize',18),
      set(gca,'ztick',[]), grid off, drawnow
    end
    vvnew = 2*vv - vvold + dt^2*(vv*D2x +D2y*vv);
    vvold = vv; vv = vvnew;
    vv([1 Ny+1],:) = BC*vv(2:Ny,:);         % Neumann BCs for |y|=1
  end
```

13.2. The lifetime $T(\epsilon)$ of the metastable state of Output 32 depends strongly on the diffusion constant ϵ. Modify Program 32 to measure $T(\epsilon)$, which can be defined precisely as the value of t at which the function $u(x,t)$ first becomes monotonic in x. (How can you test for monotonicity slickly in MATLAB?) Now perform a numerical study of $T(\epsilon)$. Find a set of axes on which to plot your results that clarifies the asymptotic behavior of $T(\epsilon)$ as $\epsilon \to 0$ and allows you to make quantitative estimates of it.

13.3. Modify Program 19 (p. 82) to solve $u_{tt} = u_{xx}$ as before but with initial and boundary conditions

$$u(x,0) = 0, \quad u_x(-1,t) = 0, \quad u(1,t) = \sin(10t).$$

Output 37

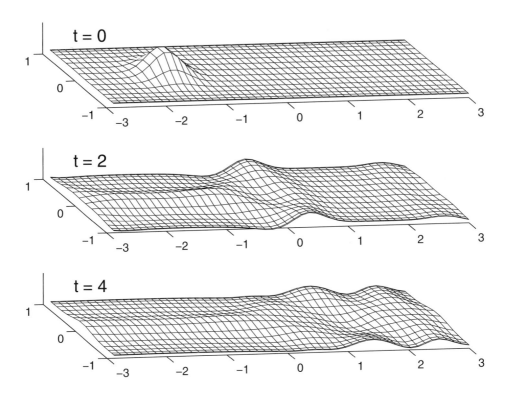

Output 37: *Solution of the wave tank problem (13.7)–(13.9) with Neumann boundary conditions along the sides.*

Produce an attractive plot of the solution for $0 \leq t \leq 5$.

13.4. The time step in Program 37 is specified by $\Delta t = 5/(N_x + N_y^2)$. Study this discretization theoretically and, if you like, numerically, and decide: Is this the right choice? Can you derive a more precise stability limit on Δt?

14. Fourth-Order Problems

The differential equations that arise in practice are usually of first, second, third, or fourth order. In this chapter we consider some fourth-order problems, both because they are interesting in their own right and because this will give us further practice with the kinds of manipulations of differential operators and boundary conditions that arise ubiquitously in spectral methods.

We begin with a one-dimensional example. Suppose that we wish to solve the biharmonic problem

$$u_{xxxx} = f(x), \quad -1 < x < 1, \quad u(\pm 1) = u_x(\pm 1) = 0.$$

Physically, $u(x)$ might represent the transverse displacement of a beam subject to a force $f(x)$. The conditions at $x = \pm 1$ are known as *clamped boundary conditions*, corresponding to holding both the position and the slope of a beam fixed at the ends.

How shall we compute the spectral approximation to u_{xxxx}? Our standard design philosophy gives an answer. Let $\{v_j\}$ be the $(N-1)$-vector of values of u sampled at $x_1, \ldots, x_N$. The "method (I)" strategy of the last chapter for imposing boundary conditions suggests the following:

- Let p be the unique polynomial of degree $\leq N+2$ with $p(\pm 1) = p_x(\pm 1) = 0$ and $p(x_j) = v_j$, $j = 1, \ldots, N-1$.
- Set $w_j = p_{xxxx}(x_j)$.

Now one might think that implementing this procedure will necessitate some new mathematics to derive the interpolant p, since the boundary conditions are different from those we have used since Chapter 6. However, we can obtain w as a by-product of our usual Chebyshev differentiation matrix D_N if we set

$$p(x) = (1 - x^2)q(x),$$

from which after four differentiations we obtain

$$p_{xxxx}(x) = (1 - x^2)q_{xxxx}(x) - 8x q_{xxx}(x) - 12q_{xx}(x).$$

A polynomial q of degree $\leq N$ with $q(\pm 1) = 0$ corresponds to a polynomial p of degree $\leq N + 2$ with $p(\pm 1) = p_x(\pm 1) = 0$. Thus we can carry out the required spectral differentiation like this:

- Let q be the unique polynomial of degree $\leq N$ with $q(\pm 1) = 0$ and $q(x_j) = v_j/(1 - x_j^2)$, $j = 1, \ldots, N - 1$.
- Set $w_j = (1 - x_j^2)q_{xxxx}(x_j) - 8x_j q_{xxx}(x_j) - 12q_{xx}(x_j)$.

At the matrix level, let $\widetilde{D}_N^2$, $\widetilde{D}_N^3$, and $\widetilde{D}_N^4$ be the matrices obtained by taking the indicated powers of D_N and stripping away the first and last rows and columns. Then our spectral biharmonic operator is

$$L = \left[\mathrm{diag}(1 - x_j^2)\widetilde{D}_N^4 - 8\,\mathrm{diag}(x_j)\widetilde{D}_N^3 - 12\widetilde{D}_N^2\right] \times \mathrm{diag}(1/(1 - x_j^2)),$$

where j runs from 1 to $N - 1$. Solving our original problem spectrally is now just a matter of solving a linear system of equations for v,

$$Lv = f,$$

where $f = (f_1, \ldots, f_{N-1})^T$.

Program 38 uses this approach to solve

$$u_{xxxx}(x) = e^x, \quad u(\pm 1) = u_x(\pm 1) = 0, \quad -1 < x < 1. \tag{14.1}$$

From the maximum error listed it is evident that the solution is spectrally accurate. Even with N reduced to 5, the error reported by Program 38 remains less than 10^{-5}.

For a more interesting result, consider the biharmonic operator in two dimensions,

$$\Delta^2 u = u_{xxxx} + 2u_{xxyy} + u_{yyyy}.$$

To discretize this operator we can employ Kronecker products as usual. For an illustration, consider the clamped plate eigenvalue problem on the unit square,

$$\Delta^2 u = \lambda u, \quad -1 < x, y < 1, \quad u = u_n = 0 \text{ on the boundary,}$$

14. Fourth-Order Problems

Program 38

```
% p38.m - solve u_xxxx = exp(x), u(-1)=u(1)=u'(-1)=u'(1)=0
%         (compare p13.m)

% Construct discrete biharmonic operator:
  N = 15; [D,x] = cheb(N);
  S = diag([0; 1 ./(1-x(2:N).^2); 0]);
  D4 = (diag(1-x.^2)*D^4 - 8*diag(x)*D^3 - 12*D^2)*S;
  D4 = D4(2:N,2:N);

% Solve boundary-value problem and plot result:
  f = exp(x(2:N)); u = D4\f; u = [0;u;0];
  clf, subplot('position',[.1 .4 .8 .5])
  plot(x,u,'.','markersize',16)
  axis([-1 1 -.01 .06]), grid on
  xx = (-1:.01:1)';
  uu = (1-xx.^2).*polyval(polyfit(x,S*u,N),xx);
  line(xx,uu)

% Determine exact solution and print maximum error:
  A = [1 -1 1 -1; 0 1 -2 3; 1 1 1 1; 0 1 2 3];
  V = vander(xx); V = V(:,end:-1:end-3);
  c = A\exp([-1 -1 1 1]'); exact = exp(xx) - V*c;
  title(['max err = ' num2str(norm(uu-exact,inf))],'fontsize',12)
```

Output 38

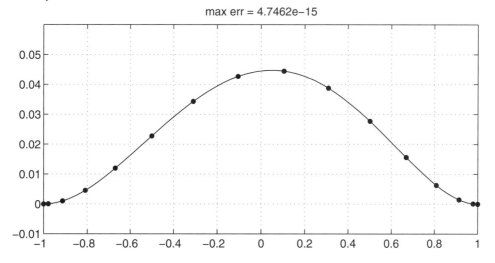

Output 38: *Solution of the one-dimensional biharmonic equation* (14.1). *We get* 14 *digits of accuracy with* $N = 15$.

Program 39

```
% p39.m - eigenmodes of biharmonic on a square with clamped BCs
%         (compare p38.m)

% Construct spectral approximation to biharmonic operator:
  N = 17; [D,x] = cheb(N); D2 = D^2; D2 = D2(2:N,2:N);
  S = diag([0; 1 ./(1-x(2:N).^2); 0]);
  D4 = (diag(1-x.^2)*D^4 - 8*diag(x)*D^3 - 12*D^2)*S;
  D4 = D4(2:N,2:N); I = eye(N-1);
  L = kron(I,D4) + kron(D4,I) + 2*kron(D2,I)*kron(I,D2);

% Find and plot 25 eigenmodes:
  [V,Lam] = eig(-L); Lam = -real(diag(Lam));
  [Lam,ii] = sort(Lam); ii = ii(1:25); V = real(V(:,ii));
  Lam = sqrt(Lam/Lam(1));
  [xx,yy] = meshgrid(x,x);
  [xxx,yyy] = meshgrid(-1:.01:1,-1:.01:1);
  [ay,ax] = meshgrid(.8:-.2:0,0:.16:.64);
  sq = [1+1i -1+1i -1-1i 1-1i 1+1i]; clf
  for i = 1:25
    uu = zeros(N+1,N+1); uu(2:N,2:N) = reshape(V(:,i),N-1,N-1);
    subplot('position',[ax(i) ay(i) .16 .2]), plot(sq)
    uuu = interp2(xx,yy,uu,xxx,yyy,'cubic');
    hold on, contour(xxx,yyy,uuu,[0 0]), axis square
    axis (1.25*[-1 1 -1 1]), axis off, colormap([0 0 0])
    text(-.3,1.15,num2str(Lam(i)),'fontsize',7)
  end
```

where u_n denotes the normal derivative. Program 39 carries out the discretization. In 20 lines of MATLAB code, we have generated images of the first 25 eigenmodes like those that can be found in many books on acoustics, mechanics, and applied mathematics.

The use of spectral methods to solve eigenvalue problems for biharmonic operators to very high accuracy is beautifully illustrated by Bjørstad and Tjøstheim in [BjTj99]. Working on the square of side length 1, without our normalization Lam = sqrt(Lam/Lam(1)), they obtain the lowest eigenvalue 1294.93397959171280817030264797944; this figure should be divided by 16 for comparison with Program 39 since the side length there is 2. With such high accuracy, they were able to resolve five regions of sign oscillation in the eigenmodes in the corners, each about 16.57 times smaller than the last; the first of these "Moffatt vortices" is visible in the plots of Output 39.

14. Fourth-Order Problems

Output 39

Output 39: *Nodal lines of the first* 25 *eigenmodes (with multiplicity) of a clamped square plate. As in Output* 28b *(p. 122), there are a number of degenerate pairs of eigenmodes: six of them, so that only* 19 *distinct eigenvalues are represented. In each degenerate pair, the eigenmodes depicted are arbitrary, and not even orthogonal, which explains why these particular modes are neither symmetrical nor familiar. Still they are correct.*

For a final, historically important example of a fourth-order spectral calculation, consider the Orr–Sommerfeld equation from the field of hydrodynamic stability [ScHe00]. (This example is adapted, with thanks, from Weideman and Reddy [WeRe00].) For flow driven by a steady pressure gradient between two infinite flat plates at Reynolds number (nondimensionalized speed) R, one solution of the Navier–Stokes equations is the laminar flow, consisting of a smooth horizontal motion with velocity profile given by a parabola:

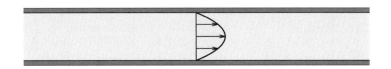

For $R < 1000$, this is the flow one observes in the laboratory. For $R \gg 1000$, however, one invariably sees turbulence instead. (The mathematics and the phenomenology are much the same for pipes as for channels.) To explain this, it is natural to expect that for high R, the laminar flow must be unstable to small perturbations. A linearized stability analysis, based on the assumption of a perturbation with longitudinal structure e^{ix} growing at the rate $e^{\lambda t}$, gives rise to the Orr–Sommerfeld eigenvalue problem:

$$R^{-1}(u_{xxxx} - 2u_{xx} + u) - 2iu - i(1-x^2)(u_{xx} - u) = \lambda(u_{xx} - u) \quad (14.2)$$

with boundary conditions $u(\pm 1) = u_x(\pm 1) = 0$. (The most sensitive perturbation actually has a dependence on x closer to $e^{(1.02)ix}$, so (14.2) is a slight simplification.) To test for instability, we look for values of R which give eigenvalues λ with positive real part. Discretizing in a manner suggested by Huang and Sloan [HuSl94], we obtain the generalized $(N-1) \times (N-1)$ eigenvalue problem $Av = \lambda Bv$, where

$$A = R^{-1}(D_4 - 2D_2 + I) - 2iI - i \operatorname{diag}(1 - x_j^2)(D_2 - I),$$
$$B = D_2 - I.$$

Here D_2 is our usual second derivative matrix $\tilde{D}_N^2$ imposing the boundary conditions $u(\pm 1) = 0$, and D_4 is the fourth derivative matrix imposing the clamped boundary conditions $u(\pm 1) = u_x(\pm 1) = 0$, as in Programs 38 and 39.

Program 40 implements these formulas to calculate eigenvalues corresponding to $R = 5772$, the critical Reynolds number determined by Orszag in 1971. The resulting eigenvalues, plotted in Output 40, converge spectrally to a Y shape familiar to all those who work in hydrodynamic stability [DrRe81]. As expected, the rightmost eigenvalue is nearly on the imaginary axis.

As it happens, the "critical Reynolds number" $R = 5772$ is not as important physically as it seemed in the 1970s, but that is another story [ScHe00, TTRD93].

Summary of This Chapter. Fourth-order problems typically require two boundary conditions at each point. In the case of clamped boundary conditions on rectangles, these can be imposed by a simple trick involving polynomials related by $p(x) = (1-x^2)q(x)$.

14. Fourth-Order Problems

Program 40

```
% p40.m - eigenvalues of Orr-Sommerfeld operator (compare p38.m)
  R = 5772; clf, [ay,ax] = meshgrid([.56 .04],[.1 .5]);
  for N = 40:20:100

    % 2nd- and 4th-order differentiation matrices:
    [D,x] = cheb(N); D2 = D^2; D2 = D2(2:N,2:N);
    S = diag([0; 1 ./(1-x(2:N).^2); 0]);
    D4 = (diag(1-x.^2)*D^4 - 8*diag(x)*D^3 - 12*D^2)*S;
    D4 = D4(2:N,2:N);

    % Orr-Sommerfeld operators A,B and generalized eigenvalues:
    I = eye(N-1);
    A = (D4-2*D2+I)/R - 2i*I - 1i*diag(1-x(2:N).^2)*(D2-I);
    B = D2-I;
    ee = eig(A,B);
    i = N/20-1; subplot('position',[ax(i) ay(i) .38 .38])
    plot(ee,'.','markersize',12)
    grid on, axis([-.8 .2 -1 0]), axis square
    title(['N = ' int2str(N) '    \lambda_{max} = ' ...
        num2str(max(real(ee)),'%15.11f')]), drawnow
  end
```

Exercises

14.1. Determine the first five eigenvalues of the problem

$$u_{xxxx} + u_{xxx} = \lambda u_{xx}, \qquad u(\pm 2) = u_x(\pm 2) = 0, \quad -2 < x < 2,$$

and plot the corresponding eigenvectors.

14.2. As noted in the caption, the degenerate eigenmodes of Output 39 did not come out symmetric in x or y because they are essentially arbitrary samples from a two-dimensional linear space. Modify Program 39 to change this. Specifically, for each degenerate pair, your program should plot one mode that is zero at $x = 0$, $y = 1$, and another linearly independent mode that is zero at $x = 1$, $y = 0$. Or does it look better if the control points are $x = 1$, $y = 1$ and $x = -1$, $y = 1$?

14.3. Another approach to symmetry is to devise a program that only computes solutions with certain symmetry properties. Construct a modification of Program 39 that works on the domain $[0,1]^2$ and imposes homogeneous Neumann boundary conditions at $x = 0$ and $y = 0$. Produce a display of the first 12 eigenmodes labeled with normalized eigenvalues as in Output 39. Which modes of that computation do your 12 modes correspond to?

Output 40

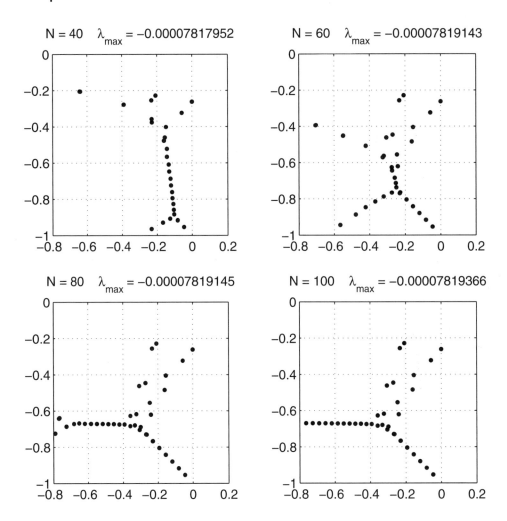

Output 40: *Eigenvalues in the complex plane of the Orr–Sommerfeld problem (14.2), $R = 5772$. For small N, the rightmost eigenvalues are accurate, and as N increases, the accuracy extends to eigenvalues deeper in the left halfplane. With this choice of R, the rightmost eigenvalue lies approximately on the imaginary axis.*

14.4. The Orr–Sommerfeld operator of Program 40 is highly nonnormal. Modify the program to produce a plot of the ϵ-pseudospectra of an appropriate spectral discretization matrix with $N = 100$ for $\epsilon = 10^{-1}, 10^{-2}, \ldots, 10^{-8}$. (See [RSH93].)

Afterword

The emphasis in this book has been on solving problems. My overriding aim has been to show you that elementary tools of spectral methods can be applied easily to get high-accuracy results in all kinds of areas.

Of course, such a style deemphasizes certain things, and one, regrettably, is mathematical depth. In two ways, this book gives too little indication of the mathematical maturity that the field of spectral methods has achieved today. One is that there has been little mention of theoretical developments, even though by now, a quarter century into the "modern" era, some remarkable mathematics has been worked out to establish the foundations of spectral methods. Numerous papers and books by Bernardi, Canuto, Funaro, Gottlieb, Maday, Quarteroni, and Tadmor, to mention just a few, represent this strong tradition. The crucial contribution of this theory, for method after method applied to problem after problem, is a proof of rapid convergence as $N \to \infty$. Spectral methods are delicate, and plausible methods frequently fail, so results of this kind are much more than confirmations of the obvious.

At the algorithmic level, this book has also oversimplified. For example, we have squared differentiation matrices by explicit matrix multiplication, so that everything can be based on the simple code `cheb`, but this is not the most stable method, nor the most efficient; the use of recurrences reduces operation counts from $O(N^3)$ to $O(N^2)$ [Wel97, WeRe00]. Similarly, we have not hinted at various more robust methods of treating boundary conditions, especially for time-dependent problems [For90, Hes00, HuSl94], or said anything about filtering or de-aliasing or other methods of suppressing linear and nonlinear numerical artifacts [CHQZ88].

And, as has been mentioned several times, the whole book is about collocation methods, not Galerkin, which is a major limitation in scope.

So please remember, this book is just a beginning! Much, much more is known about spectral methods than appears in these pages; if you don't see it here, don't assume it hasn't been done.

Here are some further examples of topics that might have been considered—and many of them *are* considered in [Boy00], [CHQZ88], [For96], or [KaSh99]. Differentiation matrices are often constructed from sets $\{x_j\}$ other than Chebyshev points [Lak86, WeRe00, Wel97]; we gave a hint of this in Exercise 6.1. The sometimes excessive clustering of Chebyshev points can be ameliorated by nonpolynomial changes of variables, and this may allow larger time steps [For88, KoTa93]. Unbounded domains can be treated by better methods than truncation to $[-L, L]$ [WeCl90, Wei95], and there is also the whole related subject of sinc function methods [Ste93]. For applications to unbounded domains, and for various other applications, some spectral methods make use of rational functions rather than polynomials [Wei95]. Many ODEs and PDEs have solutions with boundary layers, and this causes special problems for which special methods have been devised. Fluid mechanics is a very large field, and spectral methods have been utilized in every corner of it. Fluids simulations involving shock waves raise the problem of dealing with discontinuities, which appear in other kinds of problems too; there is a substantial literature on this, and spectral methods often do better than one might expect [GeTa99, GoSh98, MMO78]. Large-scale problems in all areas tend to require special methods of linear algebra, and many have been developed, often involving finite element or finite difference preconditioners of spectral discretizations. In particular, there has been much attention to spectral methods for high-performance architectures [FiPa94, KaSh99]. There is the whole matter of software, of which the MATLAB Differentiation Suite by Weideman and Reddy and the Fortran/C package PseudoPack by Costa and Don are noteworthy general-purpose examples [CoDo00, WeRe00]. There's the big, important topic of spectral methods for engineering applications, and in particular, there are the crucial techniques of domain decomposition, spectral elements, and hp finite elements [BaSu94, BeMa92, Fis97, KaSh99]. We could have talked about spherical coordinates, systems of differential equations, staggered grids, spurious eigenvalues, spectral multigrid methods, radiation boundary conditions, integral equations....

According to *Science Citation Index,* the 1988 monograph on spectral methods by Canuto, Hussaini, Quarteroni, and Zang [CHQZ88] has been cited in print at least 930 times. Spectral methods are now mainstream tools of applied mathematics.

Bibliography

[AbSt65] M. Abramowitz and I. A. Stegun, eds., *Handbook of Mathematical Functions, with Formulas, Graphs, and Mathematical Tables*, Dover, New York, 1965.

[Ahl79] L. V. Ahlfors, *Complex Analysis: An Introduction to the Theory of Analytic Functions of One Complex Variable*, 3rd ed., McGraw-Hill, New York, 1979.

[BaSu94] I. Babuška and M. Suri, *The p and h-p versions of the finite element method, basic principles and properties*, SIAM Review 36 (1994), 578–632.

[BaBe00] R. Baltensperger and J.-P. Berrut, *The errors in calculating the pseudospectral differentiation matrices for Chebyshev-Gauss-Lobatto points*, Comput. Math. Appl. 37 (1999), 41–48.

[BCM94] A. Bayliss, A. Class, and B. J. Matkowsky, *Roundoff error in computing derivatives using the Chebyshev differentiation matrix*, J. Comput. Phys. 116 (1994), 380–383.

[BeOr78] C. M. Bender and S. A. Orszag, *Advanced Mathematical Methods for Scientists and Engineers*, McGraw-Hill, New York, 1978.

[BeMa92] C. Bernardi and Y. Maday, *Approximations Spectrales de Problèmes aux Limites Elliptiques*, Springer-Verlag, Berlin, 1992.

[BjTj99] P. E. Bjørstad and B. P. Tjøstheim, *High precision solutions of two fourth order eigenvalue problems*, Computing 63 (1999), 97–107.

[Boo78] C. de Boor, *A Practical Guide to Splines*, Springer-Verlag, New York, 1978.

[Boy00] J. P. Boyd, *Chebyshev and Fourier Spectral Methods*, 2nd ed., Dover, New York, 2000.

[BrHe95] W. L. Briggs and V. E. Henson, *The DFT: An Owner's Manual for the Discrete Fourier Transform*, SIAM, Philadelphia, 1995.

[But87] J. C. Butcher, *The Numerical Analysis of Ordinary Differential Equations:*

Runge–Kutta and General Linear Methods, Wiley, New York, 1987.

[CHQZ88] C. Canuto, M. Y. Hussaini, A. Quarteroni, and T. A. Zang, *Spectral Methods in Fluid Dynamics*, Springer-Verlag, Berlin, 1988.

[CaGo96] M. H. Carpenter and D. Gottlieb, *Spectral methods on arbitrary grids*, J. Comput. Phys. 129 (1996), 74–86.

[ChKe85] T. F. Chan and T. Kerkhoven, *Fourier methods with extended stability intervals for the Korteweg–De Vries equation*, SIAM J. Numer. Anal. 22 (1985), 441–454.

[ClCu60] C. W. Clenshaw and A. R. Curtis, *A method for numerical integration on an automatic computer*, Numer. Math. 2 (1960), 197–205.

[CoDo00] B. Costa and W. S. Don, *PseudoPack, a software library for numerical differentiation*, software package, available at http://www.cfm.brown.edu/people/wsdon/home.html.

[Dav99] E. B. Davies, *Pseudospectra, the harmonic oscillator and complex resonances*, Proc. Roy. Soc. London Ser. A 455 (1999), 585–599.

[DaRa84] P. J. Davis and P. Rabinowitz, *Methods of Numerical Integration*, 2nd ed., Academic Press, New York, 1984.

[Dem97] J. W. Demmel, *Applied Numerical Linear Algebra*, SIAM, Philadelphia, 1997.

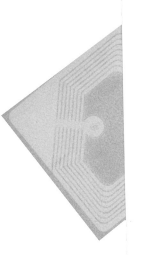

[DrJo89] P. G. Drazin and R. S. Johnson, *Solitons: An Introduction*, Cambridge University Press, Cambridge, UK, 1989.

[DrRe81] P. G. Drazin and W. H. Reid, *Hydrodynamic Stability*, Cambridge University Press, Cambridge, UK, 1981.

[Dri96] T. A. Driscoll, *Algorithm 765: A MATLAB toolbox for Schwarz–Christoffel mapping*, ACM Trans. Math. Software 22 (1996), 168–186. Software available at http://www.math.udel.edu/~driscoll/SC.

[DyMc86] H. Dym and H. P. McKean, *Fourier Series and Integrals*, Academic Press, New York, 1986.

[EmTr99] M. Embree and L. N. Trefethen, *Green's functions for multiply connected domains via conformal mapping*, SIAM Review 41 (1999), 721–744.

[Exn83] P. Exner, *Complex-potential description of the damped harmonic oscillator*, J. Math. Phys. 24 (1983), 1129–1135.

[Fis97] P. F. Fischer, *An overlapping Schwarz method for spectral element solution of the incompressible Navier–Stokes equations*, J. Comput. Phys. 133 (1997), 84–101.

[FiPa94] P. F. Fischer and A. T. Patera, *Parallel simulation of viscous incompressible flows*, Ann. Rev. Fluid Mech. 26 (1994), 483–527.

[For75] B. Fornberg, *On a Fourier method for the integration of hyperbolic equations*, SIAM J. Numer. Anal. 12 (1975), 509–528.

[For88] B. Fornberg, *Generation of finite difference formulas on arbitrarily spaced grids*, Math. Comp. 51 (1988), 699–706.

[For90] B. Fornberg, *An improved pseudospectral method for initial-boundary value problems*, J. Comput. Phys. 91 (1990), 381–397.

[For95] B. Fornberg, *A pseudospectral approach for polar and spherical geometries*, SIAM J. Sci. Comput. 16 (1995), 1071–1081.

[For96] B. Fornberg, *A Practical Guide to Pseudospectral Methods*, Cambridge University Press, Cambridge, UK, 1996.

[FoMe97] B. Fornberg and D. Merrill, *Comparison of finite difference- and pseudospectral methods for convective flow over a sphere*, Geophys. Res. Lett. 24 (1997), 3245–3248.

[FoSl94] B. Fornberg and D. M. Sloan, *A review of pseudospectral methods for solving partial differential equations*, Acta Numer. 3 (1994), 203–267.

[FoWh78] B. Fornberg and G. B. Whitham, *A numerical and theoretical study of certain nonlinear wave phenomena*, Philos. Trans. Roy. Soc. London Ser. A 289 (1978), 373–404.

[FoPa68] L. Fox and I. B. Parker, *Chebyshev Polynomials in Numerical Analysis*, Oxford University Press, Oxford, UK, 1968.

[Fun92] D. Funaro, *Polynomial Approximation of Differential Equations*, Springer-Verlag, New York, 1992.

[Fun97] D. Funaro, *Spectral Elements for Transport-Dominated Equations*, Springer-Verlag, New York, 1997.

[GeTa99] A. Gelb and E. Tadmor, *Detection of edges in spectral data*, Appl. Comput. Harmon. Anal. 7 (1999), 101–135.

[GMS92] J. R. Gilbert, C. Moler, and R. Schreiber, *Sparse matrices in MATLAB: Design and implementation*, SIAM J. Matrix Anal. Appl. 13 (1992), 333–356.

[GoVa96] G. H. Golub and C. F. Van Loan, *Matrix Computations*, 3rd ed., Johns Hopkins University Press, Baltimore, 1996.

[GoWe69] G. H. Golub and J. H. Welsch, *Calculation of Gauss quadrature rules*, Math. Comp. 23 (1969), 221–230.

[GHO84] D. Gottlieb, M. Y. Hussaini, and S. A. Orszag, *Introduction: Theory and Applications of Spectral Methods,* in R. G. Voigt, D. Gottlieb, and M. Y. Hussaini, eds., Spectral Methods for Partial Differential Equations, SIAM, Philadelphia, 1984.

[GoLu83a] D. Gottlieb and L. Lustman, *The Dufort–Frankel Chebyshev method for parabolic initial boundary value problems*, Comput. Fluids 11 (1983), 107–120.

[GoLu83b] D. Gottlieb and L. Lustman, *The spectrum of the Chebyshev collocation operator for the heat equation*, SIAM J. Numer. Anal. 20 (1983), 909–921.

[GoOr77] D. Gottlieb and S. A. Orszag, *Numerical Analysis of Spectral Methods: Theory and Applications*, SIAM, Philadelphia, 1977.

[GoSh98] D. Gottlieb and C.-W. Shu, *On the Gibbs phenomenon and its resolution*, SIAM Review 39 (1998), 644–668.

[HaWa96] E. Hairer and G. Wanner, *Solving Ordinary Differential Equations II*, Springer-Verlag, Berlin, 1996.

[Hal74] P. R. Halmos, *A Hilbert Space Problem Book*, 2nd ed., Springer-Verlag, New York, 1974.

[HJB85] M. T. Heideman, D. H. Johnson, and C. S. Burrus, *Gauss and the history of the Fast Fourier Transform*, Arch. Hist. Exact Sci. 34 (1985), 265–277.

[Hen74] P. Henrici, *Applied and Computational Complex Analysis, v. 1*, Wiley, New York, 1974.

[Hen79] P. Henrici, *Fast Fourier methods in computational complex analysis*, SIAM Review 21 (1979), 481–527.

[Hen82] P. Henrici, *Essentials of Numerical Analysis with Pocket Calculator Demonstrations*, Wiley, New York, 1982.

[Hen86] P. Henrici, *Applied and Computational Complex Analysis, v. 3*, Wiley, New York, 1986.

[Hes00] J. S. Hesthaven, *Spectral penalty methods,* Appl. Numer. Math. 34 (2000), to appear.

[Hig85] J. R. Higgins, *Five short stories about the cardinal series*, Bull. Amer. Math. Soc. 12 (1985), 45–89.

[HiHi00] D. J. Higham and N. J. Higham, MATLAB *Guide*, SIAM, Philadelphia, 2000.

[Hig96] N. J. Higham, *Accuracy and Stability of Numerical Algorithms*, SIAM, Philadelphia, 1996.

[Hil62] E. Hille, *Analytic Function Theory*, Chelsea, New York, 1962.

[HoJo91] R. A. Horn and C. R. Johnson, *Topics in Matrix Analysis*, Cambridge University Press, Cambridge, UK, 1991.

[HuSl94] W.-Z. Huang and D. M. Sloan, *The pseudospectral method for solving differential eigenvalue equations*, J. Comput. Phys. 111 (1994), 399–409.

[Ise96] A. Iserles, *A First Course in the Numerical Analysis of Differential Equations*, Cambridge University Press, Cambridge, UK, 1996.

[KaSh99] G. E. Karniadakis and S. J. Sherwin, *Spectral/hp Element Methods for CFD*, Oxford University Press, Oxford, UK, 1999.

[Kat76] Y. Katznelson, *An Introduction to Harmonic Analysis*, 2nd ed., Dover, New York, 1976.

[Kör90] T. W. Körner, *Fourier Analysis*, Cambridge University Press, Cambridge, UK, 1990.

[KoTa93] D. Kosloff and H. Tal-Ezer, *A modified Chebyshev pseudospectral method with an $O(N^{-1})$ time step restriction*, J. Comput. Phys. 104 (1993), 457–469.

[KrWu93] H.-O. Kreiss and L. Wu, *On the stability definition of difference approximations for the initial boundary value problem*, Appl. Numer. Math. 12 (1993), 213–227.

[KrUe98] A. R. Krommer and C. W. Ueberhuber, *Computational Integration*, SIAM, Philadelphia, 1998.

[Lak86] W. D. Lakin, *Differentiation matrices for arbitrarily spaced grid points*, Internat. J. Numer. Methods Engrg. 23 (1986), 209–218.

[Lam91] J. D. Lambert, *Numerical Methods for Ordinary Differential Systems: The Initial Value Problem*, Wiley, New York, 1991.

[Lan38] C. Lanczos, *Trigonometric interpolation of empirical and analytical functions*, J. Math. Phys. 17 (1938), 123–199.

[Lan56] C. Lanczos, *Applied Analysis*, Prentice-Hall, Englewood Cliffs, NJ, 1956.

[LiLo97] E. H. Lieb and M. Loss, *Analysis*, AMS, Providence, RI, 1997.

[MMO78] A. Majda, J. McDonough, and S. Osher, *The Fourier method for nonsmooth initial data*, Math. Comp. 30 (1978), 1041–1081.

[McRo00] R. I. McLachlan and N. Robidoux, *Antisymmetry, pseudospectral methods, and conservative PDEs*, Proc. EQUADIFF '99, World Scientific, Singapore, to appear.

[Mei67] G. Meinardus, *Approximation of Functions: Theory and Numerical Methods*, Springer-Verlag, New York, 1967.

[Mer89] B. Mercier, *An Introduction to the Numerical Analysis of Spectral Methods*, Springer-Verlag, New York, 1989.

[MiTa99] P. A. Milewski and E. G. Tabak, *A pseudo-spectral procedure for the solution of nonlinear wave equations with examples from free-surface flows*, SIAM J. Sci. Comput. 21 (1999), 1102–1114.

[MoIn86] P. M. Morse and K. U. Ingard, *Theoretical Acoustics*, Princeton University Press, Princeton, NJ, 1986.

[OpSc89] A. V. Oppenheim and R. W. Schafer, *Discrete-Time Signal Processing*, Prentice-Hall, Englewood Cliffs, NJ, 1989.

[Ors71] S. A. Orszag, *Accurate solution of the Orr–Sommerfeld stability equation*, J. Fluid Mech. 50 (1971), 689–703.

[OrPa72] S. A. Orszag and G. S. Patterson, Jr., *Numerical simulation of three dimensional homogeneous isotropic turbulence*, Phys. Rev. Lett. 28 (1972), 76–79.

[PaWe34] R. E. A. C. Paley and N. Wiener, *Fourier transforms in the complex domain*, Amer. Math. Soc. Colloq. Publ., v. 19, Providence, RI, 1934.

[Pey86] R. Peyret, *Introduction to Spectral Methods*, Von Karman Institute, Rhode-St.-Genèse, Belgium, 1986.

[RSH93] S. C. Reddy, P. J. Schmid, and D. S. Henningson, *Pseudospectra of the Orr–Sommerfeld operator*, SIAM J. Appl. Math. 53 (1993), 15–47.

[ReTr92] S. C. Reddy and L. N. Trefethen, *Stability of the method of lines*, Numer. Math. 62 (1992), 234–267.

[ReWe99] S. C. Reddy and J. A. C. Weideman, *The accuracy of the Chebyshev differencing method for analytic functions*, unpublished manuscript, 1999.

[RiMo67] R. D. Richtmyer and K. W. Morton, *Difference Methods for Initial-Value Problems*, 2nd ed., Interscience, New York, 1967.

[ScHe00] P. J. Schmid and D. S. Henningson, *Stability and Transition in Shear Flows*, Springer-Verlag, Berlin, to appear.

[Sol92] A. Solomonoff, *A fast algorithm for spectral differentiation*, J. Comput. Phys. 98 (1992), 174–177.

[Ste93] F. Stenger, *Numerical Methods Based on Sinc and Analytic Functions*, Springer-Verlag, New York, 1993.

[Str91] G. Strang, *Calculus*, Wellesley-Cambridge Press, Cambridge, MA, 1991.

[Tad86] E. Tadmor, *The exponential accuracy of Fourier and Chebyshev differencing methods*, SIAM J. Numer. Anal. 23 (1986), 1–10.

[Tre82] L. N. Trefethen, *Group velocity in finite difference schemes*, SIAM Review 24 (1982), 113–136.

[Tre97] L. N. Trefethen, *Pseudospectra of linear operators*, SIAM Review 39 (1997), 383–406.

[Tre99] L. N. Trefethen, *Computation of pseudospectra*, Acta Numer. 9 (1999), 247–295.

[TrBa97] L. N. Trefethen and D. Bau, III, *Numerical Linear Algebra*, SIAM, Philadelphia, 1997.

[TTRD93] L. N. Trefethen, A. E. Trefethen, S. C. Reddy, and T. A. Driscoll, *Hydrodynamic stability without eigenvalues*, Science 261 (1993), 578–584.

[TrTr87] L. N. Trefethen and M. R. Trummer, *An instability phenomenon in spectral methods*, SIAM J. Numer. Anal. 24 (1987), 1008–1023.

[Tsu59] M. Tsuji, *Potential Theory in Modern Function Theory*, Dover, New York, 1959.

[Van90] H. Vandeven, *On the eigenvalues of second-order spectral differentiation operators*, Comput. Methods Appl. Mech. Engrg. 80 (1990), 313–318.

[Van92] C. F. Van Loan, *Computational Frameworks for the Fast Fourier Transform*, SIAM, Philadelphia, 1992.

[Wei95] J. A. C. Weideman, *Computing the Hilbert transform on the real line*, Math. Comp. 64 (1995), 745–762.

[WeCl90] J. A. C. Weideman and A. Cloot, *Spectral methods and mappings for evolution equations on the infinite line*, Comput. Methods Appl. Mech. Engrg. 80 (1990), 467–481.

[WeRe00] J. A. C. Weideman and S. C. Reddy, *A MATLAB differentiation matrix suite*, ACM Trans. Math. Software, to appear.

[WeTr88] J. A. C. Weideman and L. N. Trefethen, *The eigenvalues of second-order spectral differentiation matrices*, SIAM J. Numer. Anal. 25 (1988), 1279–1298.

[Wel97] B. D. Welfert, *Generation of pseudospectral differentiation matrices I*, SIAM J. Numer. Anal. 34 (1997), 1640–1657.

[Whi74] G. B. Whitham, *Linear and Nonlinear Waves*, Wiley, New York, 1974.

[Whi15] E. T. Whittaker, *On the functions which are represented by the expansions of the interpolation theory*, Proc. Roy. Soc. Edinburgh 35 (1915), 181–194.

[Wid65] H. Widom, *Toeplitz matrices,* in I. I. Hirschman, Jr., ed., Studies in Real and Complex Analysis, Math. Assoc. Amer., Washington, DC, 1965.

[Wri00] T. G. Wright, MATLAB codes and graphical user interfaces for computation of pseudospectra, available at http://www.comlab.ox.ac.uk/oucl/work/nick.trefethen.

[Zie89] W. P. Ziemer, *Weakly Differentiable Functions*, Springer-Verlag, New York, 1989.

Index

Abramowitz and Stegun, 88–89
Adams–Bashforth formula, 52, 104–106, 113
Adams–Moulton formula, 104–106
Airy equation, 88–92
aliasing, x, 10–11, 16, 18
    formula, 32, 37
Allen–Cahn equation, 137, 139, 141
analytic
    continuation, 132
    function, 11–12, 29–30
annulus, 112
Arnoldi iteration, 97
asymptotic convergence factor, 49, 134

B-spline, 31
backward differentiation formula, 104–106
band-limited interpolation, 11–14, 19, 21, 27
barycentric interpolation, 63, 74
Bessel equation, 99
biharmonic
    equation in 1D, 146–147
    equation in 2D, 146–149
    operator, 146

bistable equation, *see* Allen–Cahn equation
Bjørstad, Petter, 148
boundary conditions
    advanced methods, 153
    clamped, 145, 150
    Dirichlet, homogeneous, 61–74
    Dirichlet, inhomogeneous, 135–141
    Neumann, 62, 99, 137–138, 140, 143–144, 151
    radiation, 154
    time-dependent, 137, 140–141, 143
boundary value problem, 61–74
bounded variation, 30, 33–34
Burgers equation, 113

C (programming language), xi, 154
Canuto, Hussaini, Quarteroni, and Zang, x, 154
cardinal function, 54, 58
CFL condition, 113
chaos, 114
Chebyshev
    density, 45
    differentiation matrices, 51–59, 109–110

differentiation via FFT, 75–86
  grids, xi, 41–50, 113
  points, 42–43, 51, 134
  polynomial, 75–77, 85–86
  series, x, 75–86
  spectral methods, 5
circulant matrix, 2–3, 7–8
Clenshaw–Curtis quadrature, xiv, 126–129, 132, 134
collocation, x, 5, 7, 154
complex
  arithmetic, 97–98, 111–112, 150–152
  FFT, 27
  plane, 29, 47, 50, 96, 98, 104, 106, 133, 152
conformal mapping, 50, 132
contour integral, 39, 132–133
convergence
  geometric, 17, 44, 59
  rates, 4
  theorems, 33–34, 48, 153
convolution, 7, 31
cubic spline interpolation, 63, 70–74, 83, 93, 142, 148
cylindrical coordinates, 115

Davies, Brian, 96
DCT, *see* discrete cosine transform
de-aliasing, 153
delta function, 12, 54
  periodic, 19
DFT, *see* discrete Fourier transform
differentiation matrices, 3
  arbitrary points, 52, 58–59, 134, 154
  Chebyshev, 51–59, 109–110
  computation by recurrences, 61, 153
  finite difference, 2–3
  Fourier or periodic, 5, 17–28
  MATLAB Suite, xiii, 154
  symmetries of, 55, 59, 124
  unbounded grid, 5, 9–16, 154
discrete
  cosine transform (DCT), 79, 85
  Fourier transform (DFT), 17, 18
domain decomposition, 154
Driscoll, Toby, xiii, 50

eigenmodes
  degenerate, 122, 149, 151
  nodal lines of, 73, 94–95, 121–122, 149
eigenvalue problem, 36, 66, 87–99
  generalized, 88–91
  perturbed, 90, 92–95
eigenvalues, nonphysical, 66, 108–110
`eigs`, 87, 99
EISPACK, xiii
ellipse, 131
elliptic integral, 131
Embree, Mark, xiv
equation
  Airy, 88–92
  Allen–Cahn, 137, 139, 141
  Bessel, 99
  biharmonic in 1D, 146–147
  biharmonic in 2D, 146–149
  bistable, *see* Allen–Cahn
  Burgers, 113
  Korteweg–de Vries (KdV), 17, 108–113
  Kuramoto–Sivashinsky, 114
  Laplace, 90, 137, 142
  Mathieu, 88–89
  Navier–Stokes, 150
  nonlinear, 63, 65, 72, 108–114
  Poisson in 1D, 61, 64–65, 135–138
  Poisson in 2D, 67, 70–71
  reaction-diffusion, 137
  variable coefficient wave, 24, 26, 101, 107, 113
  wave in 1D, 80, 82, 101, 113
  wave in 2D, 83–84, 101
equilibrium charge distribution, 49, 59
equipotential curves, 47
Euler formula, 139, 141

fast Fourier transform (FFT), 7, 23
  and Chebyshev differentiation, 75–86
  complex, 24, 27
  history of, 23
  speed of, 27
filtering, 153
finite differences, ix, 1–4, 27

finite elements, ix, xi, 154
floating point arithmetic, 7, 103
fluid mechanics, ix–x, 149–152, 154
Fornberg, Bengt, xiii, 110, 115
Fortran, xi, 154
Fourier
    analysis and synthesis, 9–10
    differentiation matrices, 5, 17–28
    differentiation via FFT, 25
    Joseph, 23
    series, 11, 75
    spectral methods, 5
    synthesis, 10
Fourier transform, x, 9
    decay of, 29
    fast (FFT), 7, 23
    inverse, 10
    semidiscrete, 10–11
    symmetries of, 16, 24, 79
fourth-order problems, 145–152
function
    analytic, 11–12, 29–30
    bounded variation, 30, 33–34
    cardinal, 54, 58
    delta, 12, 54
    delta, periodic, 19
    gamma, 132–134
    Green's, 49–50
    harmonic, 43
    rational, 154
    sinc, 13–14
    sinc, periodic, 20–21

Galerkin spectral methods, xi, 130, 132, 135, 154
gamma function, 132–134
Gauss
    Carl Friedrich, 23
    quadrature, 42, 126–132
geometric convergence, 17, 44, 59
Gibbs phenomenon, 14–15, 41
Gottlieb, David, x, 108
Green's function, 49–50
grid
    clustering of, x, 41–50, 67–68, 84–85, 113, 115–117, 154
    tensor product, 67, 84

    transformed, 42, 154

Handle Graphics, xvi
harmonic
    function, 43
    oscillator, 36, 38, 87, 96–98
    oscillator, complex, 96–98
Helmholtz equation, 69, 72–73
Hermite polynomial, 36, 96
Higham brothers, xiv, xvi
high-performance architectures, 154
high-precision calculations, 148
holomorphic, see analytic
hp finite elements, xi, 154

IEEE arithmetic, 7
integral equations, 154
integrating factors, method of, 111–112, 113
integration, see quadrature
interpolation, 2, 7, 51
    band-limited, 12–14, 19, 21, 27
    barycentric, 63, 74
    cubic spline, 63, 70–74, 83, 93, 142, 148
    in Chebyshev points, 77
    in equispaced points, 77
    in roots of unity, 77
    of spectral results, 62–63, 122, see also cubic spline interpolation and polynomial interpolation
    polynomial, 42, 48, 62–65, 71
    sinc function, see band-limited interpolation
    trigonometric, 5, 19, 130

Korteweg–de Vries (KdV) equation, 17, 108–113
Kronecker
    product, 68, 72, 92, 97, 118, 146
    sum, 69
Krylov subspace iteration, 97
Kuramoto–Sivashinsky equation, 114

$L^2(\mathbb{R})$, 10
$L^2([-\pi/h, \pi/h])$, 11
$\ell^2(\mathbb{Z})$, 11

Lanczos
    Cornelius, x
    iteration, 97
Laplace
    equation, 90, 137, 142
    operator in polar coordinates, 117–119
    operator in 2D, 69–71, 92
    operator in 3D, 99
LaTeX, xiv, xvi
Laurent
    operator, 5
    polynomial, 76, 126
    series, x, 75, 131
    series and FFT, 132
leap frog
    formula, 24–28, 80–83, 101, 104–107
    formula, second order, 107, 140, 143
Legendre points, xi, 129, 134
lexicographic ordering, 68, 92
LINPACK, xiii

machine epsilon, 7
Mathieu equation, 88–89
MathWorks, ix, xiii
MATLAB, ix–xiii
membrane, circular, 88, 97, 99, 117–122
memory requirements, x
metastability, 137, 143
method of lines, 103
Moffatt vortices, 148–149
Moiré patterns, 10
Moler, Cleve, xiii
multigrid, 154

Navier–Stokes equations, 150
Newton's method, 74
nilpotent matrix, 59, 99
nonlinear equation, 63, 65, 72, 108–114
normal and nonnormal, 96, 104, 152

O and o, 29
ODE, ix, 24
operation count, 24, 27, 59, 61, 153

Orr–Sommerfeld problem, 97, 149–152
Orszag, Steve, x, 87, 150

Paley–Wiener theorems, 30
PDE, ix, 24
periodic
    boundary conditions, 17
    domain, 17, 88, 130–131
plate, square, 97, 146–149
points per wavelength (ppw), 66
Poisson
    equation in 1D, 61, 64–65, 135–138
    equation in polar coordinates, 123
    equation in 2D, 67, 70–71
    summation formula, 32
polar coordinates, 17, 115–124
polynomial
    Chebyshev, 75–77, 85–86
    Hermite, 36, 96
    interpolation, 42, 48, 63, 71
    minimax, 49–50, 97
    trigonometric, 76
potential theory, x, 43–50
PseudoPack, 154
pseudospectra, 87, 96–99, 104, 113, 152
    vs. pseudospectral, 96
pseudospectral, *see* collocation

quadrature, 125–134
    Clenshaw–Curtis, 126–129, 132, 134
    Gauss, 42, 126–132
    trapezoid rule, 11, 131–134
quantum mechanics, 36
QZ algorithm, 90

rational function, 154
reaction-diffusion equation, 137
resonance, 72
Reynolds number, 87, 149–152
Riemann zeta function, 16
Riesz representation theorem, 126
rounding errors, xiv, 6–7, 27, 49, 55, 59, 85, 103
Runge–Kutta formula, 24, 101–106
Runge phenomenon, 42–45

sampling theorem, 12
sawtooth mode, 20, 102, 107

self-reciprocal, 76, 126
semidiscrete Fourier transform, 10–11
Schwarz–Christoffel
    mapping, 50
    MATLAB Toolbox, xiii, 50
Science Citation Index, 154
Shannon, Claude, 12
Signal Processing Toolbox, 79, 86
sinc function, 13–14
    interpolation, *see* band-limited interpolation
    methods, 154
    periodic, 20–21
singular value decomposition, 97
solitons, 17, 109, 112
sparse matrix, 2–4, 69, 71, 74, 99
spectral
    accuracy, x, 6–7, 29–39, 41, 58, 85, 127–130
    elements, xi, 154
    methods, history of, ix–xi
    multigrid, 154
spherical coordinates, 17, 115, 117, 154
Spratley, Andrew, xiv
stability
    hydrodynamic, 87, 149–152
    numerical, 74, 101–114
    region, 101–114
    restrictions on Δt, 101–115, 144
stroboscopy, 10
superellipse, 134
symmetry, exploitation of, 59, 117–119, 124, 134, 151

tau methods, 135
Taylor series, 1, 8, 50, 131
    and FFT, 132, 134
tensor product, 68
    grid, 67, 84
time-stepping, 101–114
Toeplitz matrix, 2, 5, 13
trapezoid rule, 11, 131–134
trigonometric
    interpolation, 5, 19, 130
    polynomial, 76
    series, x
Trefethen, Emma and Jacob, xiv

turbulence, 17, 150

unbounded domain, 9, 154

Vandermonde system, 134

wagon wheel effect, 10
wave
    equation in 1D, 80, 82, 101
    equation in 2D, 83–84, 101
    equation, variable coefficient, 24, 26, 101, 107, 113
    nonlinear, xiii, 108–114
    tank, 140, 144
wavenumber, 10
Weideman, André, xiii, 88, 149
Whittaker, Sir Edmund, 12